D0679613

FOAL

THE
LONG
AND THE
SH⚙RT
OF IT

THE
LONG
—AND THE—
SHRT
OF IT

HOW WE CAME TO
MEASURE OUR WORLD

GRAEME
DONALD

Michael O'Mara Books Limited

First published in Great Britain in 2016 by
Michael O'Mara Books Limited
9 Lion Yard
Tremadoc Road
London SW4 7NQ

Copyright © Michael O'Mara Books Limited 2016

All rights reserved. You may not copy, store, distribute, transmit, reproduce or
otherwise make available this publication (or any part of it) in any form, or by
any means (electronic, digital, optical, mechanical, photocopying, recording or
otherwise), without the prior written permission of the publisher. Any person
who does any unauthorized act in relation to this publication may be liable to
criminal prosecution and civil claims for damages.

A CIP catalogue record for this book is available from the British Library.

Papers used by Michael O'Mara Books Limited are natural, recyclable products
made from wood grown in sustainable forests. The manufacturing processes
conform to the environmental regulations of the country of origin.

ISBN: 978-1-78243-628-7 in hardback print format
ISBN: 978-1-78243-623-2 in ebook format

1 2 3 4 5 6 7 8 9 10

Designed and typeset by K DESIGN, Winscombe, Somerset

Printed and bound by CPI Group (UK) Ltd, Croydon, CR0 4YY

www.mombooks.com

CONTENTS

For my ever-patient wife, Rhona, whose love and support over the past years defies measurement.

INTRODUCTION

In order to make sense of our own environment and that of the universe beyond, we need agreed scales of measurements which allow us to quantify the tangible and the intangible. Whether you are baking a cake or estimating the fuel requirements of a jumbo jet embarking on a transatlantic flight, you need a precise understanding of the volume of the raw materials involved and to be able to factor in variables of weight, time and temperature. You need measurements.

It wasn't until people abandoned nomadic lifestyles for settled communities that the need arose to measure. They had to measure out individual plots of land, which gave rise to basic geometry (from Greek *ge*, land, and *-metria*, a measuring of), and they needed to weigh out the grain they grew for sale or barter. The human body provided a ready, common scale, so horses were measured in hands (and still are, although the hand has been standardized to 4 inches), while different gauges were made by other body parts, such as the thumb, foot, forearm from fingertip to elbow and the span of outstretched arms. Each early settlement must have used a rock or stone as a grain weight – hence the imperial weight unit of a stone – but as this would have varied in size, standards to codify scales of weight and volume had to be introduced. As commerce expanded and found itself in need of employees, time itself had to be measured to make sure the labour force turned up on time and worked the agreed number of hours.

As with many other basic features of everyday life, such as language, regional codes of measurement developed into national ones, as it became essential to understand what vendors were offering and to have an appreciation of size and value. It may strike some readers that the following pages are biased towards the imperial and metric scales, but these are the measurements by which most of the world now runs. With the possible exception of the ancient Malay unit *kati* (600 g / 1 ⅓ lb), which gave us the tea caddy, few exotic and obsolete measurements would even be recognized in their country of origin, so their inclusion in this book could hardly be justified.

Other current units – no pun intended – such as the volt, amp and ohm have also been excluded because, after the complexity of their definition is explained, there is little of interest to say about them. Much the same reasoning was behind the exclusion of units such as the pascal, the erg, the dyne and the atmosphere, which are more familiar in the fields of science and engineering than in everyday life.

The origins of many standard measurements are intriguing. The boxing ring became square and has the dimensions it does today because of the length of the four-horse coaching whip, while the shape and dimensions of the modern tennis court originate in the old manorial courtyard. Tennis' somewhat eccentric scoring derives from the face of the old courtyard clock. But why are clock faces traditionally round? And what determined clockwise rotation? The answers lie in the sundial: had mechanical clocks been invented in the southern hemisphere, then clockwise would denote a rotation to the left.

European shoe sizes – 38 or 40, for example – equate to the number of barleycorns laid end to end, while the mile is based on one thousand (*mille*) full marching paces of a Roman legionary. The format of the periodic table was based on that of the card game solitaire, or patience. Zero degrees on the original Fahrenheit scale was determined by the temperature of ice in Danzig (now Gdańsk, Poland) during the unusually severe winter of 1708–9, while its upper marker of one hundred degrees was the temperature in Frau Fahrenheit's armpit.

It is probably fair to say that we all take measurements for granted; a yard is a yard, a kilogram is a kilogram and so what? But as with so many other seemingly mundane subjects, when you delve into the history and origins of measurements you are in for more than a few surprises.

HEIGHT, LENGTH
AND DEPTH

Measuring the height of anything is entirely dependent on which baseline you use. When it comes to mountains, it is customary to record the height in feet or metres above sea level, which is all well and good, but specifies neither the body of water being taken as the benchmark, nor the time of year. The surface height of all bodies of water changes quite significantly by the hour, due to tides, and from week to week due to shifts in planetary and lunar orbits. Nova Scotia's Minas Basin, for example, has a tidal range of twelve to fifteen metres or forty to fifty feet, as has the Severn Estuary, which lies between England and South Wales.

To solve this problem, the scientific collective opted for mean sea level as measured at the midpoint between high and low tides — but, again, different nations took their readings from different bodies of water.

In 1915, the UK fixed the mean sea level as the mid-tide point in Cornwall's Newlyn Harbour, which is fine for measuring peaks and promontories throughout the UK but not much use for measuring those in other lands. When you see a height claiming to be an accurate measure in feet or metres above sea level, you have to ask which sea was taken as the baseline and in which month that mean level was established. This is necessary even if the figures quoted are based on satellite readings, as all seas are in a permanent state of flux.

YOU *CAN* FEEL THE EARTH MOVE!

Most people imagine that tides are an exclusively oceanic phenomenon for which the moon alone is responsible, but the sun too plays a significant part with its own gravitational pull. Combined, these two heavenly bodies have no trouble in pulling ripples on the land masses.

Known as earth tides, these ripples can lift the surface by as much as 55 cm / 22 in. at the equator, with about 15 cm / 6 in. of that figure attributable to solar pull. Opinion in the relevant scientific sectors is divided as to the degree to which these ripples are responsible for seismic and volcanic activity.

MOUNT EVEREST

Arguably the world's highest mountain, this was known by Victorians as Peak XV. As far as they were aware, the now iconic mountain was a modest promontory with its own base over 17,000 ft / 5,182 m up the Tibetan Plateau. From its base to its peak, Mount Everest is only about 12,000 ft / 3,658 m, but it has the unfair advantage of sitting atop the Himalayan massif. The tallest free-standing mountain is Africa's Mount Kilimanjaro at 19,340 ft / 5,895 m.

Everest (properly pronounced 'Eve-rest', after Sir George Everest (1790–1866), one-time surveyor general of India) was first measured in 1852, by the Indian mathematician Radhanath Sikdar (1813–70), a senior figure of the Great Trigonometric Survey under its director, Sir Andrew Waugh (1810–78). When Sikdar reported to Waugh that Peak XV was exactly 29,000 ft high, Waugh was so worried that such a figure would invite accusations of a lazy rounding-off that he delayed publication. Now humorously hailed as the first man to put two feet on the top of Everest, when Waugh did release the figure in 1856, he announced the peak to be 29,002 ft high.

A more accurate figure was determined by a US expedition of 1999, members of which swept away all the snow from the peak to hold a GPS unit to the tip of the rock to get a reading of 29,035 ft (8,850 m).

At least, that is the latest accurate measurement. As the Himalayas are still rising out of the ground by as much as one inch per year, with the entire range moving steadily northwards into China, come the next millennium the tip of that peak will be about eighty feet, or twenty-four metres, higher.

THE HAIR'S BREADTH

Now relegated to the realm of metaphor, the smallest measure of width ever in general use was the human hair's breadth. Once common to cultures ranging from ancient Hebrew to medieval English, there were forty-eight hair's breadths to the inch. Used until the turn of the twentieth century by jewellers and watchmakers, the hair's breadth enjoyed a brief revival during the Second World War (1939–45), when it was used in the cross hairs of snipers' telescopic sights and bombsights.

Although Asian hair is in many ways much stronger than Caucasian, it is also more than twice the diameter and thus less suited to the fine workings of optical reticules. With the war having cut off supplies of long blonde hair from Scandinavia, in 1943 the US military ran a series of newspaper advertisements asking for women with blonde hair in excess of 22 in. (56 cm) in length to come forward – but only if their hair had never been weakened by dyes and bleaches or been subjected to the rigours of hot irons. This narrowed the field to the point that only one applicant stepped forward: Mary Babnik Brown (1907–91) of Colorado, a dancer whose 34 in.- / 86 cm-long tresses had only been washed with baby shampoo.

Having truthfully been told that her hair was needed to manufacture meteorological instruments, Mary's head was shaved and she resorted to a bandana made from a small Stars and Stripes. But Mary's hair was also being used to produce the reticules, or cross hairs, in the early versions of the top-secret Norden bombsight. Although it is widely maintained that Mary's hair was in the sights of both the *Enola Gay* and *Bockscar* aircraft when they bombed Hiroshima and Nagasaki in 1945, this is not true as, by then, all Norden reticules were etched into the glass viewer with diamond-tipped tools. Be that as it may, in 1987 President Reagan wrote to Mary on her eightieth birthday to tell her how her hair had helped win the war.

THE INCH

Most early measurements were based on parts of the body or other natural items that tend to be consistent in size, and the inch embraces both of these concepts. The word is derived from the Latin *uncia*, meaning one twelfth part, which in this case is one twelfth of a foot, cognate with an ounce, which was fixed at one twelfth of a troy pound: troy being the system of weights employed in the French town of Troyes, about 150 km / 93 mi. southeast of Paris, and once famous for its precious metals markets.

With the width of the average male adult thumb at the middle knuckle approximating one inch (2.54 cm), this became widespread as a rough measure throughout most of Europe and Scandinavia, which explains why in French, Spanish, Portuguese, Dutch, Swedish, Norwegian, Danish, Czech, Hungarian and Slovak the words for thumb and inch are pretty much the same.

The size of thumbs can vary from person to person so, after the Norman invasion of 1066, William the Conqueror (1028– 87) fixed the inch as the length of three barleycorns lying end to end. This was later enshrined in a statute of 1324, imposed by Edward II (1284–1327), which defined the inch as being the length of 'three grains of barley, dried and round, placed end to end, lengthwise'. The only echo of this definition to resound in modern life is in shoe sizes, which Edward also decreed should be defined by the barleycorn. Although the UK has since moved on to shoe sizes of, say, 6, 6 ½ or 7, the old barleycorn-based scale survived in Europe to be reimported, so that shoes in most UK retailers show a size of, for example, 7 and 40, to indicate that seven is the equivalent in length of forty barleycorns laid end to end.

THE SCOTTISH INCH

With the English inch sorted out and legally defined, confusion persisted in Scotland where King David I (1084–1153) had decreed in his mid-twelfth-century Assize of Weights and Measures that the inch should be set at the width of a man's thumb at the base of the nail.

When it was pointed out to David that this concept would be subject to the vagaries of individual variance, he ushered into the court a host of men of various builds. Their thumbs were then measured to produce an average, setting the Scottish inch at a length equating to 1.0016 of the English inch.

This might not seem much to argue about but, until the Act of Union of 1707 did away with David's rogue inch, there was many a disagreement between Scottish and English cloth traders over deals involving significant shipments of finished cloth, as that 0.0016 variance racked up to several yards. Such squabbles resulted in Scottish vendors saying of English buyers 'give them an inch and they'll take an ell': an expression in which the ell has now been updated to a mile. The ell was an old measure which, like the cubit (see page 28), represented the distance from the elbow to the tip of the middle finger.

RULE OF THUMB

There is a belief dating back to the eighteenth century that 'rule of thumb' refers to the legal entitlement of men to beat their wives with a stick no thicker than their thumbs, yet no evidence of this can be found in legislature.

More prosaically, the expression relates to the use of the thumb by carpenters and the like, as a rough guide to the inch. The thumb is also used in planting, as both a measure between seeds and the depth to which they should be pushed. Artists have long used the thumb of an outstretched arm as a means of establishing perspective with a handy gauge of distance.

Your arm measures ten times the distance between your eyes so, with one eye shut, hold out your arm with the thumb vertical and align it with a distant building of known size, say a 100-ft barn. Holding still, shut the first eye and open the other to see the thumb jump to one side. If it appears to jump five barns' width, that gives you a figure of 500 ft. Multiply this by ten (the ratio of arm to eye) to get a final figure of 5,000 ft, which places you about a mile from said barn.

THE SWEDISH INCH

In Sweden in 1863, their inch, called the *tum* (thumb) and initially 2.47 cm against the English inch's 2.54 cm, was suddenly, for no apparent reason, expanded to 2.96 cm. Some stubborn Swedes refused to adapt, others observed the change, and those working in machining and carpentry adopted the English inch as their standard, which only increased the confusion.

Even after Sweden started its slow and painful ten-year transition to metrication in 1878, such professionals continued to use the English foot and inch, and did so until the early twentieth century. This is why, even today, many things in Sweden from boats to television sets, are routinely measured in feet and inches. Textiles, too, are still graded in Sweden by threads per square inch, and Swedish McDonald's outlets continue to promote their Quarter Pounder – some things you just don't mess with!

OTHER INCHES

Problems arose for the first European traders dealing with India, where the Indus inch equated to 1.32 English inches (3.35 cm). Either the denizens of said region had very big thumbs, or they were defining their inch by multiples of rice grains, the latter being more likely.

Similar problems were encountered in China where, instead of using the width of the thumb as a guideline, they measured their inch from the tip to the inside crease of the middle knuckle, equating to 1.312 English inches (3.3 cm). In fact, until 1799 when the French, keen to establish a uniform standard that transcended political borders, came up with the metric system that was so eagerly embraced by most nations, the length of an inch depended very much on where you were.

Naturally, these problems were increasingly exaggerated as you moved up the measurement scale to longer units based on multiples of whichever inch you were working with. In pre-metric France the inch was 1.06 of an English inch (2.69 cm). This variance gave rise to the myth that Napoleon I (1769–1821) was vertically challenged. His height was given as 5 ft 2 in. (1.57 m) on the imperial French scale, with few on the British side of the Channel realizing at the time that this equated to nearly 5 ft 7 in. (1.7 m). People were then generally shorter than now. Lord Nelson, for example, was only 5 ft 5 in. (1.65 m), while the average Frenchman stood about 5 ft 4 in. (1.62 m) tall. But of course, Napoleon wasn't French: born Napoleone di Buonaparte, with the last element of his surname pronounced like 'party', he was Corsican-Italian.

Napoleon Bonaparte

THE BOAT'S ON THE OTHER FOOT

In most cases, variance in the length of the feet and inches used by different countries served only to irritate and confuse, but in seventeenth-century Sweden it actually sank a warship and killed the crew.

In 1628 Sweden launched what was reckoned to be the most powerful warship of her day, the 64-gun *Vasa*, but the crowd's cheers turned to cries of dismay as she sank but 1,300 m / 0.8 mi. from shore, with the loss of many lives. When she was raised in 1961 the problem was identified: she was thicker and longer on the port side than she was on the starboard side, a fatal lack of symmetry that brought her down.

Some of the shipwrights' tools were still on board and it seems that the Swedish team working on the port side were using their country's pre-1863 feet and inches, while the Dutch workforce on the starboard side were working to their own Amsterdam foot, a measure just short of eleven Swedish inches.

AIR AND RAIN

Atmospheric pressure is still measured in inches of mercury, as determined by the barometer designed in 1643 by the Italian Evangelista Torricelli (1608–47).

A simple calibrated tube standing upside down in a reservoir of mercury, this is still reckoned to be the most accurate of all barometers, with increased pressure on the reservoir of mercury forcing ever more of the liquid metal up the tube to give a higher reading. One atmosphere at sea level is equivalent to 14.7 lb/in.^{2}, which will register 29.93 in. of mercury.

Rainfall, too, is measured in inches, but such figures are grossly misleading to the general mind. The greatest rainfall during a twenty-four-hour period in recent times was the 73.62 in. (187 cm) that fell on La Réunion on 15 and 16 March 1952, but this does not mean that the island was swamped with over six feet (1.8 m) of water.

The standard rain gauge comprises a collection funnel that feeds into a calibrated tube that is one-tenth of the funnel's diameter. This ten to one ratio means that even the lightest showers can be harvested in the funnel to produce a readable figure in the collection tube – but one that must always be divided by ten.

This leaves the deluge that hit La Réunion registering a less impressive but more realistic figure of 7.36 in. (18.7 cm) of rain which, across the island's 970 mi^2 (2,512 km^2), still adds up to a serious amount of rainwater.

THE MYTHICAL INCH

Inspired by the groundless speculations of others before him – including the British physicist and mathematician Sir Isaac Newton (1643–1727), who was obsessed with alchemy and the paranormal – Charles Piazzi Smyth (1819–1900), Astronomer Royal of Scotland, spent much of the 1870s in Egypt, trying to prove that the pyramids were astronomical devices constructed to reflect the dimensions of the earth and the distances between certain heavenly bodies.

Smyth wrongly inferred from Biblical reference to the alleged presence of the Hebrew slaves that the Sacred Hebrew Cubit must have been used as a prime unit in the construction. Well, that was his first mistake – there was not a single Israelite, slave or otherwise, involved in the construction of the pyramids of Egypt; the labour force was entirely Egyptian.

That aside, Smyth set to work, determined to prove the pyramids to be more than just the very large piles of stones they are. He measured every aspect of the structures after deciding that the hitherto undefined Sacred Cubit of the Hebrews must have comprised twenty-five sacred inches, each being the equivalent of 1.00106 standard British ones. Using this cart-before-the-horse reasoning to make sense of all the measurements he had already taken meant that, to Smyth at least, all became clear.

The original perimeter of the Great Pyramid, as determined by Smyth himself, measured 36,524.2 sacred pyramid inches, or one hundred times the number of solar days in the year. Dividing by twenty-five the number of sacred inches in the height of any one side of the Great Pyramid to get back to the number of Sacred Cubits, the answer was also 365.242. Smyth then realized that the Polar circumference of the earth was exactly 250,000,000 of his sacred pyramid inches.

There was much more such hokum, and there are some people who still subscribe to his theories of the pyramids as sacred instruments, built to measure the universe. No matter that in 1883 Sir William Flinders Petrie (1853–1942), the renowned Egyptologist, established beyond any doubt that Smyth's measurements were wildly inaccurate and that the Great Pyramid was in fact several feet shorter than he maintained, all of which made a nonsense of Smyth's Sacred Cubit and inch, the pyramidologists' bandwagon is still rolling.

THE FOOT

Only since July 1959 has it been internationally agreed that the foot shall be a fixed measure equating to 0.3048 m and that it comprise twelve inches, each equating to 25.4 mm – before this, chaos reigned.

Historically, the measurement equated to the length of a human foot, usually that of the ruler or monarch of the time, but with the foot of the average male representing about a sixth of his height, the length of the foot varied from country to country, depending on the average height of the race. The ancient Greeks and Romans both used the foot as a basic measure with the former having the larger feet, so their foot equated to 302 mm while the Roman foot was smaller at 295.7 mm, or about 97 per cent of the current standard measurement.

At least, that was the standard foot within the city of Rome. Out in the Roman provinces most favoured the larger *pes Drusianus* which, at 334 mm, was actually in use long before the advent of Nero Claudius Drusus (38–9 BC). Just to confuse matters further, within the City of Rome all temples and civic buildings were constructed to the dictates of a sacred foot of 325 mm.

The early Romans and Greeks both subdivided their foot into sixteen digits – each the width of the index finger – but the Romans later opted for a larger division they called the *uncia*, or one-twelfth part, whence we derive both 'inch' and 'ounce'. In 29 BC, Marcus Agrippa (63–12 BC), Roman statesman and the most senior general of the Roman Army, standardized the Roman foot to the length of his own right foot – 296 mm / 11.5 imperial in. – and it was this measurement that was ushered into the UK with the Roman invasion of AD 43.

Following Roman withdrawal in AD 409–10, the Saxons arrived to fill the void, bringing with them the northern German foot which equated to 335 mm / 13 in., and was subdivided into four hand-palms or twelve thumbs. Rather confusingly, the 'new' Saxon foot was used to measure out plots of land but the buildings erected thereon continued to be constructed to the dictates of the old Roman foot. In 1300, or thereabouts, the Composition of Yards and Perches, convened under Edward I (1239–1307), redefined the foot by using the odd concept of three barleycorns to the inch and thirty-six of the same to the foot, to promote the new statute foot that equated to ten-elevenths of the Germanic big foot.

As stated above, this remained pretty much unchanged until July 1959 when the international foot was unveiled, this being exactly 0.3048 m and 1.7 parts-per-million longer than Edward's foot.

THE BOXING RING GOES SQUARE

Back in the days of bare-knuckle fighting, the action took place within the circle of spectators who gathered in any open space or field large enough to accommodate the event. The outer ring of that circle was formed by the gentry, come to see the fun and bet heavily on the result. Sitting atop their coaches for a better view, they tired of the result being influenced by spectators either tripping up one of the fighters or distracting him with the odd jab in the ribs, so it became the custom for their burliest coachmen to be sent in to stand four-square, with their coaching whips stretched between them to hold back the mob.

The kind of whip used on a four-horse coach measured about 5.2 m / 17 ft from the base of the handle to the tip, so each side of this impromptu square comprised a whip and two coachmen's arms of about 90 cm / 3 ft each. This made for a square of 7 m / 23 ft, which is pretty much what the boxing 'ring' measures today.

CUBITS AND ELLS

Once commonly employed units of measurement, both the cubit and the ell corresponded to the distance from the elbow to the tip of the middle finger – about 46 cm / 18 in. This made a handy rule for carpenters, who could quickly assess the length of timbers and planks by using their own right arms. Although no one talks of ells today, this was in fact the width chosen for the standard computer keyboard: a symmetry designed to ease the strain of prolonged typing.

After establishing their empire, the Romans redefined the cubit as being the distance from the left hip bone to the clenched thumb and index finger of the fully extended right arm, as this made a handy way of measuring out cloth or rope. The vendor would hold the edge of the fabric or the end of the rope against his left hip so the buyer could see how many times this was drawn out by the full and upwards extension of the right arm. This redefined cubit or ell was about 114 cm / 45 in. One presumes that few Romans would opt to buy rope or cloth from a vertically challenged vendor.

It was this definition of the cubit that the Romans brought with them to Britain, which explains why the standard width of a bolt of cloth in the UK is still 114 cm / 45 in. As with all such measuring methods, the problem was that three cubits of cloth or ten cubits of rope, for example, would vary slightly from vendor to vendor and, eventually, groat-conscious Brits tired of buying from salesmen who stood measuring out like someone auditioning for *Saturday Night Fever*.

With the cubit accepted as being 45 in. (114 cm), brass tacks were hammered into the vendors' workbenches at this distance, a practice that persisted with haberdashers until the close of the nineteenth century, although by then the tacks marked out yards and inches. Either way, this is why people still insist on 'getting down to brass tacks' instead of relying on guess and conjecture.

WHATEVER FLOATS YOUR BOAT

The most famous structure laid out in the old eighteen-inch (46 cm) cubit has to be Noah's Ark, which allegedly measured three hundred cubits long, fifty cubits wide and thirty cubits high, or, 137 m / 450 ft long, 23 m / 75 ft wide and 14 m / 45 ft high.

Built, we are told, out of mysterious 'gopher wood' this would have snapped in half on the crest of the first wave. It would also have been nowhere near big enough to accommodate ninety-odd million species of insect, reptile, plant life and animals, the vast majority of which, the Bible also tells us, had to be collected seven by seven and not the two by two of popular imagination.

180!

The game of darts – and it is a game, not a sport – was invented by medieval archers messing around on the village green outside the pub. Having taken an empty keg for their game of butts, as it was called, they chalked concentric circles on the top before setting it out at the distance of one chain, to see which of them could lob a hefty iron dart and strike nearest to the centre. This game was also used to teach novice archers the art of trajectory.

In time, and given the typical English summer, the barrel top was eventually hung inside on the tavern wall, the divisions refined and the numbers shuffled to a pattern that minimized the chance of fluke scoring. But how to fix the distance of the oche? Well, in a pub, what else but four beer crates set end to end, to give a distance from the target of something in the region of 2.4 m / 8 ft. Naturally, this varied slightly from pub to pub, depending on the size of crates furnished by their different suppliers, so an average of 2.37 m / 7 ft 9.25 in. was finally established.

SHORT MEASURES

The ancient Greek version of the cubit was the *pygme*, which varied slightly in that it was taken from the elbow to the knuckles of the clenched fist. The early Greeks encountered the diminutive sub-Saharan Africans through their interaction with North African civilizations, and named them after this measure. As for the people so classified, their lives are short too: fewer than half the children live to see the age of fifteen and the average adult is dead before they are twenty-five – but the anthropological jury is still out as to why this should be.

THE YARD

This unit of linear measurement has come a long way since the seventh century when, in England under Danelaw, a yard of land or a yardland was roughly a quarter of a hide of land. With a hide being more a measure of worth than actual area, in that the richer and more productive the land then the smaller the hide, a yardland could be anything from fifteen to thirty acres (six to twelve hectares).

By medieval times the yard had shrunk to a linear measuring stick of about sixteen feet or five metres that was used in multiples to mark out acres: now more commonly referred to as a rod, this measuring stick is still used by surveyors. Henry I (*c.* 1068–1135) tried to set the yard to the length of his own right arm, but Edward I swept this way with his Composition of Yards and Perches, a committee that sat *c*. 1300 and decreed that three grains of barley, laid end to end, made an inch, twelve inches made a foot, three feet made a yard, five-and-a-half yards made a perch and that an area measuring forty by four perches made an acre.

Elizabeth I (1533–1603) seems to have been the first to commission a formal and official yardstick in 1588 and, at a mere 0.01 in. / 0.25 mm shorter than the current official yardstick, it was not a bad effort. But in 1758, a committee under Lord Carysfort decided that the Elizabethan yardstick was both badly made and bent and commissioned London instrument maker John Bird to knock up a new one. Bird made a second in 1760, but both his yardsticks were destroyed in the fire of 1834 that brought down the old Palace of Westminster.

In 1838, another committee commissioned the astronomer Francis Baily (1774–1844, of Baily's beads fame) to make a new yardstick, but he died in 1844 after passing the baton

Queen Elizabeth I

to the Reverend Richard Sheepshanks (1794–1855, grandfather of the Camden Town Group painter Walter Sickert, whom some reckon a likely candidate for having moonlighted as Jack the Ripper). Working to a tolerance of one part in ten million, Sheepshank's thirty-six-inch (91 cm) yard was signed off by Queen Victoria (1819–1901) as the new official standard for the Empire on 5 August 1855: sadly, the day after he died.

COURTS AND TENNIS

The small, rectangular courtyards of modest monastic cloisters and manors were communal areas ideal for playing real tennis, the forerunner of lawn tennis. Mainly a game for the rich, real tennis was a bit like squash in that the walls and columns were all in play for ricochet shots, giving rise to the popular idiom 'knocked from pillar to post'. Indeed, in the early eighteenth century, the once-wealthy inmates of London's debtors' prisons adapted real tennis to a game they could play in the exercise yard. Known as rackets, or prison rackets, this was the version of real tennis that would evolve into squash in the early 1830s, when it was played in yet another variant form at Harrow School.

When the upstart lawn tennis arrived in the early 1870s it was perhaps inevitable that the court would be based on the old courtyard rectangle, as favoured by real tennis with its five-by-two-rods layout. The unusual scoring of modern tennis also reflects the method of keeping score in real tennis on small wooden replicas of the courtyard clock. Originally this was 15, 30, 45 and 60 (game), but 45 was moved back to 40, the twenty-to position on the clock face, to allow for the concept of deuce. Introduced to prevent a player winning by a single point, this derived from the French *deux*, two, as each player was then required to take two consecutive points to secure the game. Sometimes the umpire's call at this point was *jeu parti*, French for 'game divided', which evolved into 'jeopardy', as each player was in danger of losing.

Wimbledon held its first championships in 1877 and as that club had, because of its own space restrictions, opted for courts measuring seventy-eight by thirty-six feet (24 m x 11 m) this became standard by 1882, when the club was already standing supreme in the game.

DISTANCE

Moving about the countryside of pre-industrial Europe was arduous by day and impossible by night, so the concept of distance tended to be indicated by the number of days it would take to walk or ride to the destination; this explains why 'travel' and 'journey' are based on the French *travail*, meaning work or labour, and *jour*, a day. But, as a day's travel by foot, horseback or coach could all account for different distances on level ground, with further variances imposed by hilly or rough terrain, larger units of measurement and a network of marker stones were established so that all travellers, whether mounted or not, could figure out for themselves how long it might take them to complete their journey.

Most of early Europe adopted the Roman mile, determined by one thousand marching paces of a legionary, and you would be forgiven for presuming that ended the confusion – not so. Depending on where you were travelling, you had to know what the host nation meant by one mile, as this could vary quite dramatically. The longest is the ten-kilometre (6 mi.) *mil*, still used in parts of Scandinavia. Perhaps this is why so many countries eagerly adopted the ten-base metric system introduced in 1799 by the First French Republic. Be that as it may, first prize for ingenuity in the measurement of long distances must go to the ancient Greek Eratosthenes, discussed below.

THE FIRST MAN TO MEASURE THE WORLD

The ancient Greek scientist Eratosthenes of Cyrene (*c.* 276–194 BC) knew that on the summer solstice the sun at noon was in its zenith, or directly overhead, in the Egyptian town of Swenet (now Aswan). He also knew it was 5,000 stadia from Swenet to Alexandria, one stadion being an ancient Greek linear measure equivalent to roughly two hundred yards or one hundred and eighty-three metres, and used to mark out sporting grounds, hence 'stadium'.

Armed with these two facts, Eratosthenes went to Alexandria to set up a sundial and await the next summer solstice, so he could take a bearing on the sun at noon above that location. The variance between the reading he had taken at noon in

Swenet (perhaps in the year 240 BC) and that taken at noon in Alexandria the following year proved to be seven degrees and twelve minutes, or one-fiftieth of a circle. Next, he calculated there to be seven hundred stadia to the degree, which meant that the variance of seven degrees and twelve minutes accounted for 5,040 stadia, which, in turn multiplied by fifty, gave him a final figure of 252,000 stadia for the circumference of the earth. This would now equate to 39,690 km / 24,662 mi. which, given the now-established circumference at the equator of 40,075 km / 24,901 mi., is a pretty fair effort for a chap working over two thousand years ago, with nothing more sophisticated than a sundial.

POLES APART

As the earth bulges at the equator and flattens off at the poles, the equator's circumference is greater than that at the poles. But there are actually fifteen poles, so where do you take the measurement?

There are two geographic poles, two magnetic poles, two geomagnetic poles, two celestial poles and, just for good measure, there is a ceremonial South Pole, about 180 m / 590 ft from the geographic South Pole. On top of that, there are four poles of inaccessibility: Northern, Southern, Oceanic and Continental.

Drifting several kilometres a year, the magnetic North Pole is now to be found in the Arctic Ocean, about 850 km / 528 mi. from the geographic North Pole, and the magnetic South Pole is located miles outside the Antarctic Circle, about 2,860 km / 1,777 mi. from the geographic South Pole. Magnetic compasses will soon be rendered redundant by a second magnetic North Pole growing in Siberia, while a second magnetic South Pole is powering up off the coast of Brazil. That should make navigation fun!

CITY LIMITS

We have all seen signs proclaiming with authority that it is, for example, one hundred miles to Manchester, eighty kilometres to Paris – but to which point in the city do they refer? In the nineteenth century most British cities other than London took the main post office as the reference point, but this has now largely been replaced by the town hall.

When it comes to capital cities, in Paris the matter is clear: there is a brass kilometre zero plaque, as the point is known, outside the front door of Notre-Dame Cathedral, and all distances to Paris relate to this specific point. In Rome, there is a trig-stone atop Capitoline Hill which serves the same purpose, and in Tokyo it is the midpoint of the Nihonbashi Bridge that serves as its kilometre zero.

Berlin leaves none in doubt as the city boasts a towering reconstructed Prussian milestone in front of the city's Spittelkolonnaden at precisely 52.510788 degrees north, 13.398964 degrees east, no less. Addis Ababa has its own kilometre zero outside St George's Cathedral, as decreed by Haile Selassie (1892–1975)

The Old Charing Cross

in 1930, but matters are clouded when it comes to London, as this is the only European capital with no agreed point zero.

Some London agencies take the site of the old Charing Cross as their reference point. This spot is now marked by the statue of Charles I (1600–49) in Trafalgar Square, which boasts a plaque stating *that* to be the point from which all distances from London are measured. Others work to the site of the London Stone, an ancient monolith in Cannon Street, while traditionalists take as their marker the threshold of St Mary-le-Bow Church, of Bow bells fame. Cities throughout the UK nod to the old Post Office system by giving distances from their own town halls to the Post Office Tower (now the BT Tower) in London's West End.

To be fair, all of the above reference points are within two or three miles or kilometres of each other but, in this day of pinpoint satnav accuracy, and given that other European cities have all made the effort, one would think the various London agencies could agree on one specific point and splash out on a brass plaque themselves.

FURLONGS, FATHOMS AND LEAGUES

Equating to two hundred and twenty yards (201 m) or one quarter of a mile, the word furlong is a corruption of the

Old English for a long furrow as ploughed down the length of an acre, which was ten chains long and one chain wide, or 220 yd long x 22 yd wide (201 m x 20 m). It was the difficulty of turning a yoke of stubborn oxen that convinced the 'yokel' to plough in that pattern, despite the drainage advantages accrued from short ploughing across the acre.

Few apart from those in the world of horse racing are now concerned with furlongs, but they were used to measure out the city blocks of, for example, Chicago and Salt Lake City, and through some bizarre hangover from British involvement in Burma (now Myanmar) road signs in that country still give distances in a combination of miles and furlongs.

The fathom takes its name from another Old English term meaning 'with outstretched arms', and was based on the distance between the fingertips of arms so held, or six feet (182 cm). As the gesture of flinging the arms apart made for a handy way to measure or coil rope, it was mainly at sea where the fathom found favour, especially as a way of expressing depth on a sounding line (the weighted line used to measure the depth of water) that had been knotted at each pause in this measuring process.

As is perhaps well known to some, Mississippi riverboats once used a sounding line with only three knots – the mark over, the mark twain and the mark under – and it was the middle marker, indicating a safe two fathoms, that was taken as a pen name by US humourist Samuel Langhorne Clemens (1835–1910).

When it comes to burial at sea, the Admiralty supposedly dictated that the body be weighted and 'buried' in no less than six fathoms of water. Not only did this give rise to people talking of 'deep-sixing' unwanted junk, but sailors' talk of 'the deep-six' confused landlubbers, who ran away with the misconception that land burials have to be in graves dug six feet deep. In fact, the UK's only stipulation states that there must be no less than two feet (61 cm) from the top of the coffin to the lip of the grave.

The league is more of a puzzle, as it did not originate as a specific distance but as the distance a person could walk in one hour: naturally, this meant that a league would vary depending on the age and level of fitness of the individual and the terrain being negotiated. That said, a league on land was generally accepted as being the equivalent of three miles (just under 5 km) and at sea three nautical miles (5.5 km).

In France a league was four kilometres (nearly 2 ½ miles) but those who delight in announcing the impossibility of the title of Jules Verne's *Twenty Thousand Leagues Under the Sea* (1870), as no sea is that deep, should be made aware that the popular English version of Verne's title is a mistaken rendition of his *Twenty Thousand Leagues Under the <u>Seas</u>*, which related to the distance travelled, not the depth attained. But, outside that title and Alfred Lord Tennyson's (1809–92) poem *The Charge of the Light Brigade* (1854) in which the troopers rode 'Half a league onward', the term is now rarely heard.

GUNTER'S CHAIN AND THE CRICKET PITCH

Both rope and string cut to a specific length have been used to measure out everything from land to stacks of wood – hence a cord of firewood, which should be 8 ft long by 4 ft wide and deep (2.4 m long x 1.2 m wide and deep). This was measured by an eight-foot (2.4 m) length of twine or cord, which could be folded in half to test the other two dimensions. But rope can stretch, and it was also an easy job for swindlers to shorten their rope and rebind the cut end. In an effort to curb such dishonest practice, in 1620 the English clergyman and polymath Edmund Gunter (1581–1626) came up with a chain comprising one hundred paper-clip-like links of 7.92 in. (201 mm) each, to give an overall length of twenty-two yards (just over 20 m).

Gunter's chain became a standard in surveying and land measurement in general, so in rural areas of Britain where the game of cricket began, it was perhaps inevitable that it would be used to lay out the village cricket pitch, which still measures twenty-two yards between stumps. The chain was also used to set up archery ranges, by measuring from the rear support of the target to the oche, where the archer stood at a range of twenty yards from the face of that target: this is still the standard distance for indoor practice.

With eighty inflexible and unstretching chains to the mile, Gunter's invention was ideal for the planning and laying out of nineteenth-century railways, a role it fulfilled right up to the planning of the High Speed 1 Channel Tunnel Rail Link, which opened its first section in 2003. The plans for that project all spoke of chainage as the prime measurement, and travellers on UK trains today can still see distance-markers displaying, for example, '134 / 63 to London' indicating there are one hundred and thirty-four miles and sixty-three chains to the capital terminus.

THE MILE

Arguably the first metric unit of measurement introduced to Britain, the mile was brought in by the ancient Romans for whom it was one thousand – or *mille*, in Latin – full marching paces of a legionary (that is, the full swing of one leg from one ground touch to the next). When marching through uncharted territory there was always one officer delegated to keep count of the pace so he could drive into the ground a specially carved stick to mark off each *mille*. At the end of the day's march, a stone marker was put in place to denote the number of miles from the starting point of the trek, and so was born the milestone.

The first domestic milestones sited by Romans stood along the Appian Way, and as their network of straight roads expanded they erected in the central Forum of the city the now-lost Golden Milestone, so that all other milestones throughout the Empire could also give the distance to that point. Not only did this produce the notion that 'all roads lead to Rome', but in 1923 it also inspired the USA to site the zero milestone in Washington DC, just south of the White House, as a reference point to which all distances in the country could relate so they, too, could make a similar boast about all roads leading to Washington.

The Roman marching-mile could vary in length due to the incline and harshness of the terrain being covered but, irked by the increasing number of short-stepped miles imposed on his military roads by the trudging steps of exhausted squaddies (as mentioned on page 35) in 29 BC Marcus Agrippa, Commander in Chief of the Roman Army, decreed that five times the length of his own foot was the standard marching pace to produce the 5,000 ft (1,524 m) Roman mile. This remained the standard for the English mile until the Weights and Measures Act of 1593 under Elizabeth I, who decided that everything should be based on the old concept of three barleycorns to the inch, twelve such inches to the foot and three such feet to the yard to give us the new mile: 5,280 ft (1,609 m).

OTHER MILES

When it came to the length of a mile outside Elizabethan England, chaos reigned. The old Welsh mile equated to three miles and 1,470 yards (6.2 km), as based on nine thousand single steps, while the Scots got closer with a mile of 1,976 imperial yards (1,806 m, the distance of Edinburgh's Royal Mile). Although these rogue miles were swept away by the Act of Union in 1707, the Irish held onto their mile of 2,240 yards (2,048 m), or 1.27 standard miles, arguing that if that was the number of pounds in a ton then it should also be the number of yards in a mile. Hard as it may be to follow that reasoning, this mile remained in common use in the Republic of Ireland until the nineteenth century and in rural regions was still enjoying sporadic use for the first quarter of the twentieth century.

Until the adoption of the metric system in the late nineteenth century, the Austro-German mile equated to 7.5 km / 4.6 mi., while the Hungarians and the Russians went for 8.3 km / 5.1 mi. and 7.4 km / 4.5 mi. respectively. Throughout Scandinavia the mile could vary from 6 to 14.5 km / 3.7 to 9 mi. and, long after such countries opted for metrication in the late 1880s, Norway, Finland and Sweden still use *mil* to describe a distance of 10 km / 6 mi. The Swedish tax agency still calculates allowances for business travel by car by the *mil*, and fuel consumption of vehicles is commonly stated in litres per *mil* in those three nations.

In her autobiographical *You May Well Ask* (1979), Lady Naomi Mitchison (1897–1999), one of the most underrated of all Scottish writers, recalled with no little chagrin her misguided willingness to join a group of Swedes for a three-*mil* walk along the coast, only to find herself embarked on a hike of truly epic proportion.

THE NAUTICAL MILE

Unlike its landlubbing cousin, the nautical mile is not a fixed distance and is closely linked to the knot, in that a ship travelling at one knot should cover a nautical mile in one hour.

If you could hold the world in one hand and cut it in half at the equator you would be left looking at a circle. If you then divided that circle into three hundred and sixty degrees and each of those degrees into sixty minutes, then each of those minutes would represent a nautical mile. This would be fine if the world were a perfect globe, but it is an oblate sphere that bulges at the equator and flattens at the poles, which in turn means that the nautical mile could vary from 6,046 ft (1,842.8 m) on the equator to 6,108 ft (1,861.7 m), depending on where you were sailing.

This might not seem much in the grand scale of things, but over a long voyage a ship could miss its destination by miles if the navigator failed to properly conduct the complex calculations required to allow for the varying length of the nautical mile as the journey progressed, and for the fact that the ship was travelling over the surface of an imperfectly shaped globe rather than a flat plane.

Help arrived in the 1580s when the Flemish cartographer Gerardus Mercator (1512–94) published a series of charts showing the world as a flat – or plane – rectangle, with all the meridians intersecting at right angles, but without distorting the navigational process. Freed from the need for the aforementioned mind-bending calculations, the navigator's job became 'plane sailing' – not plain sailing. But to achieve this new ease of navigation, Mercator's maps created quite a few misconceptions, some of which are detailed below.

The British introduced the Admiralty nautical mile, which plumped for an average of 6,080 ft (1,853 m), a compromise generally accepted by other maritime nations, but not without its own problems of inaccuracy. The Monaco Conference of 1929 eventually defined the nautical mile as being precisely 6,076.115 ft / 1,852 m, which remains the international nautical mile to this day.

As for the knot, this is exactly as it sounds. Each ship carried a rope that was knotted at intervals of 47 ft and 3 in. (14.4 m) and attached to a wooden log. On command, a man tossed the log over the side so the navigator could count the number of knots dragged out by the passage of the ship through the water in the time it took his twenty-eight-second sandglass to empty.

With the speed of a vessel being of equal importance to the known direction of travel at sea, the results were noted down in what would become known as the logbook, now a term for a ledger of any important information.

THE MISCONCEPTION OF MEASUREMENT ON MAPS

To generate the map of the world so familiar today, with the sphere of the world shown as if flat and with all the meridians intersecting at right angles, Gerardus Mercator had to progressively exaggerate the size of the landmasses furthest away from the equator and grossly distort their shape. This distortion presents many anomalies in the measurements of longitude and latitude.

No matter how it appears on the map of the UK, Edinburgh – which sits on pretty much the same longitude as Timbuktu in Mali, West Africa – is actually further west than both Bristol and Cardiff. London, at fifty-one degrees north, is further north than the main line of the US–Canadian border, which runs along the forty-nine degrees north line before dipping south after the Great Lakes. Also, if you fly due east from New York the first landmass you see is Portugal.

When it comes to size, Africa, for example, looks about the same as North America yet it is big enough to accommodate the USA, Canada, Greenland and Central America and still have enough room left for Bolivia, Chile and Argentina. The Sahara desert alone, tucked up in the top left corner of Africa, is itself about a third of a million square miles larger than the main block of the USA, Alaska excluded. Africa's offshore island nation of Madagascar may look tiny but is more than fourteen times the size of Switzerland.

METRES AND KILOMETRES

Although various luminaries of post-revolutionary France lay claim to the concept of the metre, it was in fact first promoted in 1668 by the English cleric and philosopher John Wilkins (1614–72). Wilkins championed Christopher Wren's (1632–1723) proposal that a new and exact measure was needed and that it should be based on the length of the rod required to allow a pendulum to generate the one-second action in a long-case clock. As things turned out, this was a length of 39.26 in. / 997 mm – close, but no cigar.

The next suggestion was that the new metre should be one ten-millionth of the length of a meridian that ran from the North Pole to the equator, or one forty-millionth of the polar circumference of the world. This was the suggestion favoured by the French Academy of Sciences on the grounds that the pendulum-rod option was unreliable, as the force of gravity varies slightly from country to country since the planet is not a perfect sphere. By 1873 the academy had produced what it claimed was the definitive metre bar and there the matter rested for nigh on a century.

In 1889 it was realized that the much-vaunted metre bar held at the academy was in fact two hundred micrometres short – one fifth of a millimetre – as those who had produced it had forgotten to properly account for the flattening of the planet at the North Pole. The academy did its best to keep this revelation under wraps and continued to maintain that its dodgy metre was the definitive measure. They largely got away with this until 1960, when the French General Conference on Weights and Measures decided to check the metre against – wait for it – 1,650,763.73 wavelengths of the orange-red (light) emission line in the electromagnetic spectrum of the krypton-86 atom in a vacuum.

But still this did not satisfy all quarters so, in 1983, that same French General Conference finally nailed the metre to be that distance travelled by light in a vacuum in one 299,792,458th of a second, and there the matter rests – for the time being, at least.

METRIC UK?

Although successive British governments have been discussing the adoption of the metric system since 1818, it is fair to say that this is yet to be fully embraced in the UK. Even when the UK joined the European Economic Community (later the EU) in 1973, many recalcitrant British traders persisted in selling goods by the 'old' imperial weights and measures, thereby leaving themselves open to prosecution by a Weights and Measures Board that made some capricious decisions.

Such folly reached its zenith in 2008 when Weights and Measures officers targeted Nic Davidson for selling lager by the litre in his Doncaster restaurant. He was ordered to drop the metric measure and serve by the pint but, as Davidson correctly pointed out, since EU legislation on such matters took precedence over UK rulings, the officers should back down. With help from such unlikely sources as then Prime Minster Gordon Brown, Davidson won his case.

That said, the British government itself is yet to fully comply with the EU rulings requiring dimensions on warning signs to be given in metric only or in both metric and imperial, and since the UK voted to leave the EU in June 2016, such requirements are in the balance. UK road signs still state distance in miles and not kilometres and, far more dangerous, many of the warning signs on low bridges still give height and width in feet only.

After assessing the millions of pounds' worth of damage inflicted each year on non-motorway bridges in the UK by foreign drivers trying to convert in their heads feet to metres as they approach low bridges, the administration agreed that, as of March 2015, there was to be a phased programme of sign replacement to show the clearance height and width of all bridges in both feet and metres. There will not, however, be any move to give distances in kilometres or speed limits in both miles per hour and kilometres per hour.

KLICK

As for the kilometre, there is little mystery in that it is simply the span of a thousand metres, but the now widely used synonym of 'klick' does seem to have a story behind it, even if all cannot agree on the precise origin.

What is known is that it first arose in the US military, possibly during the Korean War, but it was most certainly in broad use by US troops throughout the Vietnam War. The expression is maintained by many to have been born of the fact that there were two range adjusters on heavy mortars and larger artillery. One of these took the range of fire out in kilometre steps, with the other used for finer adjustments between such markers. It was the kilometre-step adjuster that shifted position with a clear and audible 'click', which morphed into 'klick' to keep in step with the spelling of kilometre.

GAUGING THE DISTANCE TO THE STARS

To call space travellers astronauts is a bit of an exaggeration, as none has yet reached the stars. Apart from our own sun, the next closest star to earth is Proxima Centauri, which stands 1.3 parsecs (4.24 light years) out from the sun, which is itself nearly 149.6 million km / 92.9 million mi. away.

A parsec equates to 3.26 light years / 31 trillion km / 19 trillion mi., which is a considerable distance to measure, so astronomers resort to parallax readings. As the earth orbits the sun the stars appear to shift position in the night sky, and the closer the star the greater the apparent shift. A bearing is fixed on the target star when the earth is at what we might call the nine o'clock position in its solar orbit. Six months later, by which time it has travelled 299 km / 186 million mi. to the three o'clock position on the other side of the sun, a second bearing is taken.

The difference between these two bearings – and these are very small margins indeed – gives the distance to the star. A star with a positional shift of a mere 1/2,600th of one degree, or one second of the arc, is exactly one parsec – or parallax second – beyond the sun.

Until recent technological advances, this form of measurement was only good up to one hundred parsecs but we are now talking in terms of kiloparsecs, megaparsecs and even gigaparsecs.

The centre of the Milky Way is eight kiloparsecs from earth and the entire 'swirl' a staggering thirty-four kiloparsecs in diameter. The galaxy of Andromeda is 780 kiloparsecs distant, or about 2.5 million light years, so if aliens really are paying us regular visits, they are crossing mind-boggling distances just to abduct and probe a few unfortunates.

VOLUME AND AREA

The mathematical term 'geometry' is simply the Greek for land measurement, and a skill that evolved to ensure the accurate measurement of areas of land or to settle boundary disputes, as was the case with the Mason–Dixon Line. Although most imagine this was made to mark the north-south divide of the USA, it was in fact a non-political and relatively short line established by surveyors Jeremiah Dixon and Charles Mason to settle a 1760s land dispute between the Calverts of Maryland and the Penns of Pennsylvania.

Readers will also notice that some measurements of area, regarded today as fixed units, started out as more nebulous concepts based on the area of land the average man could plough in a day and, as land quality varied from one location to another, so too did the size of the acre. As for volume, this is derived from the Latin *volvere*, to roll, and was first used for rolls of parchment. The term shifted to the books that replaced such scrolls and it was the large dimensions of early books that caused 'volume' to take on overtones of bulk and size.

THE ACRE

In medieval times the area of land a man owned and the amount of livestock it could support was the main measure of wealth, so it was essential, if only for assessing the amount of tax owed, to establish a nationally accepted system of measuring land. But unlike today, when most accepted units of area are fixed in terms of square feet, yards or miles, most medieval measurements of area related more to the amount of ploughing a man could accomplish with a team of oxen within a given time frame.

In early England it was reckoned that a fair day's work for a man with a yoke of oxen would result in him ploughing a patch of land that was forty times as long as the sixteen-and-a-half-foot (5 m) rod he used to goad the oxen, and four times that rod's width. This day's work fixed the size of the acre as being a patch of ploughed land that measured 220 yd x 22 yd (201 m x 20 m) – 4,840 yd² (4,047 m²). In times of strife, the yokels tipped their ox-goads with metal spikes to turn them into pikes and the 5.5-yd (5 m) rod is still a standard tool carried by modern surveyors. Best of all, the ploughman needed no supervision as he could tell with ease when his work was done as the answer to that question was, quite literally, in his own hands.

So, the standard acre was established by the unit of a day's work for a man presented with good, level, arable land – but what if the land to be put under the plough was impacted and dry, waterlogged and poorly drained, or if it lay at an angle? Common sense came into play to alter the size of the acre from county to county and country to country. In Ireland, for example, the acre comprised 7,840 yd^2 (6,555 m^2), while in arable Cheshire it comprised 10,240 yd^2 (8,561 m^2). In medieval Italy, where the land was not of the best quality, the acre dropped to 1,507 yd^2 (1,260 m^2) and it fell to 1,196 yd^2 (1,000 m^2) in Greece and Turkey.

In medieval Germany, erstwhile time-and-motion men realized that both man and beast would be fresher in the morning than in the afternoon so the *Tagwerk*, or day's work, was cut in two with the *Morgen* being the measure of work a ploughman would be expected to accomplish in the morning, that constituting about 65 per cent of his *Tagwerk*. But again, soil quality meant that, across Germany, the *Morgen* varied dramatically from the 2,279 yd^2 (1,905 m^2) expected in tough-to-plough Homburg (whence the famous hats) to the 6,035 yd^2 (5,046 m^2) demanded in the dairy lands of Holstein. In northern Germany's über-arable Hadeln, the Morgen expanded to an eye-watering 14,000 yd^2 (11,705 m^2).

SONS OF THE SOIL

Throughout landowning Europe, it was traditional for the oldest son to inherit the better acreage, while his younger brothers were allocated land of progressively less value. In southern Spain, where such tradition persisted until the mid-twentieth century, the least arable land was the scrubland closest to the sea, this being the unenviable lot of many a youngest son until the tourist boom of the 1960s, when the youngest sons started to sell off their valuable land to developers.

In England, this fragmentation of farms by inheritance was countered by successive Acts of Enclosure. These put farms too small to be practicable under compulsory purchase orders so the dividing hedgerows could be dug up and collective units put under new ownership and modern management. Between 1604 and 1914 there were 5,283 such acts, resulting in the 'traditional' grand patchwork of the English countryside.

THE HIDE

Next up in size from the acre was the hide, an Anglo-Saxon term meaning family, as this was the extent of land expected to furnish a livelihood for the average peasant and his family. Similar to the acre of moveable size, the hide too was more an estimation of worth or richness of the land and was adjudged as being the equivalent of eight oxgangs: an oxgang being an old

Danelaw concept based on the area that one ox could plough in the planting season. Again, this depended very much on the quality of the land but, on average, a hide comprised anything from sixty to one hundred and twenty acres (24–49 hectares).

After the Norman Conquest, the hide was redefined as that amount of land expected to yield the holder £1 of income a year. Today, £1 will buy something relatively meagre from your local pound shop, yet this was a significant sum in the twelfth century, when you could buy an ox for a shilling (five new pence). However, the holder of a hide would be taxed at that rate whether or not the land had yielded such income that year; no excuses. In fact, it is worth mentioning here that the man we now call a farmer was then called a house-bond – whence 'husband', as you could not hold a hide unless you were married – and firmer, or farmer, was the title of the tax collectors.

Tax collecting at this time was time-consuming, and if the officials did not live in the area and know who was hiding what then a shortfall could be expected. So the Crown subcontracted to men who bought the tax collection concession for a given number of hides – or indeed a whole county – for a fixed, or firm, price and then set about trying to tax a profit out of the locals (hence the idiom to 'farm out' work). In time, the title of these firmers, or farmers, shifted to men who rented a parcel of land in the hope of making a profit over and above the 'firm' rent they had paid.

HUNDREDS

Hides were gathered into divisions of a shire called a hundred as, in the eyes of the Crown and for military purposes, such an area would be expected to muster that many men under arms in times of need. With hides varying in size, depending on the quality of the land and the density of the population, a hundred also contained in practice anything from eighty to one hundred and twenty hides. Each hundred was controlled by the iron fist of a baron, put in place by the Crown to maintain the law, so this old size of his fiefdom also came to denote the size of the standard barony.

The hundreds were also gathered into larger subdivisions of a shire called a rape or a rope, from the medieval custom of law courts being held in the open but roped off from the crowd. There was also usually a hitching rail erected in front of the magistrates and anyone with something pertinent to say was 'called to the bar'. The jurisdiction of these old ropes evolved into the typical area covered by a county court, and urban district councils were created to administer the day-to-day running of the area. There was a larger division of a shire called a thridding – one-third part – but this is now only remembered in Yorkshire in the form of Riding.

The hundreds are also largely forgotten, save the legal nicety of accepting election to the Chiltern Hundreds, an option open to any British MP wishing to quit their seat in the Commons mid-term.

PINTS AND GALLONS

Both pints and gallons started out in medieval times as measures of wet or dry goods and, to a lesser extent, both still fulfil these twin functions. The first definition of the pint was the combined volume of thirty-two mouthfuls of fluid as spat out into a container that then had a line painted on the outside as the level marker: it takes its name from the Latin *pingere*, meaning painted.

In fact, the unedifying measure of the spat-out mouthful lies at the root of many imperial measures of volume, both past and present. Obviously, when it came to the larger measures listed below, these were simple multiplications of the thirty-two-mouthful pint, as no one could monitor the 32,768 mouthfuls that made up the butt – imagine losing count halfway through! In later and more fastidious times the mouthful was rebranded as the tablespoonful, but in medieval days it was ruled that there were:

2 mouthfuls to the pony
4 to the jack
8 to the gill or jill
16 to the cup
32 to the pint
64 to the quart
128 to the pottle
256 to the gallon

512 to the peck
1,024 to the kenning
2,048 to the bushel
4,096 to the strike
8,192 to the coomb
16,384 to the hogshead
32,768 to the butt

JACK AND JILL AND THE MUNCHKINS

Although the jack is largely now obsolete, the gill or jill is still sometimes used to measure spirits. The still-popular nursery rhyme might be a satire on the fact that in 1625 Charles I tried to raise a little extra revenue by downsizing the jack and the jill while maintaining the same taxes, only to have his scam blocked by parliament.

Eventually tiring of definitions dictated by some peasant's oral capacity, Elizabeth I decreed that the pint should comprise twenty fluid ounces, as indeed it still does. In America, it was decided after the War of Independence (1775–83) that it would be more sensible to fix their 'liberty' pint to that volume occupied by one pound of water. This is still the definition, which is why the US pint, at sixteen fluid ounces, is shorter than the imperial measure.

From the late medieval era until the early nineteenth century the Scottish equivalent of the pint was the mutchkin, which corresponded to three-quarters of an English pint. Through the notion of this being something of a short measure, the term was adopted in the altered form of munchkins by L. Frank Baum (1856–1919), a man of Scots-Irish heritage, in *The Wonderful Wizard of Oz* (1900).

The Scottish also had a measure they called a pint but, confusingly, this equated to three English pints. This recognized unit remained in service in Scotland until the close of the nineteenth century. With the pronounced Scottish influence in Canada, the old Canadian pint was likewise three English pints (1.7 l) – now reduced to a quart, or two pints (about 1 l) – while the equally significant French influence gave Canada the old French chopin – half a Scottish pint – for both liquids and dry goods. Just as in England, where some types of seafood are still sold by the pint, in Canada one can still buy punnets of blueberries and other small fruits packaged by the chopin.

THE CHOPIN

The Polish composer Frédéric Chopin's (1810–49) French father had an ancestor who had been a wine broker and their family name is taken from the chopin measure: once a standard retail unit of wine siphoned off the main barrel. The old French chopin was 0.852 of the yet-to-be-established litre (1 ½ pt) and was, in pre-revolutionary France, the standard size of a bottle of wine. After metrication, the chopin was downsized to a more convenient 0.75 of a litre (1 ⅓ pt) and it was this smaller chopin that was used in multiples to produce the outlandishly named champagne bottles of unmanageable size that run from the three-litre / four-chopin / five-pint Jeroboam up to the thirty-litre / forty-chopin / fifty-three-pint Melchizedek – just try lifting that to your lips!

THE LITRE

In post-revolutionary France the litre was defined as being that volume of water required to fill a cube with ten-centimetre sides. This was thought to weigh a kilogram but, as proper consideration had not been given to the purity of the water and the right temperature to ensure its maximum density, the measure was slightly under par. Nor were matters much helped by the fact that the definitive kilogram weight kept in the French Academy was itself overweight by twenty-eight parts-per-million, which left this proto-litre measuring 1.000028 l / 1 ¾ pt proper.

This remained the case until 1901 when the French General Conference on Weights and Measures decided to put right this minor error by abandoning the notion of a ten-centimetre cube and opting instead for one kilogram of pure, distilled water at its maximum density of 3.98 °C and subjected to the pressure of exactly one atmosphere.

At the twelfth sitting of that same General Conference in 1964 it was decided to redefine the litre yet again by reverting to the 1790s concept of the ten-centimetre cube, so that a litre of water now weighs a fraction less than a kilogram but is close enough for rule-of-thumb calculations. Be that as it may, 'litre' is no longer used in scientific circles, where it has been replaced by 'cubic decimetre'.

THE C-CUP

There is even a reason why the standard champagne coupe glass has a circumference of 30 cm (12 in.) and a capacity of 260 ml (9 fl oz). Formally known in France as a *bol-sein*, or breast-cup, according to some experts this style of glass was first manufactured from a mould taken from the left breast of a youthful Marie Antoinette (1755–93), who was but fourteen when she married Louis XVI (1754–93) of France in 1770. The mould was taken three years later and the glasses produced, possibly as the prank of a vengeful teenager who could have a private chuckle at the spectacle of her sworn enemy and husband's mistress, Madame du Barry (1743–93), drinking from one such at court.

Marie Antoinette

ABV AND PROOF

In the seventeenth century the minimum standard for British Navy rum was an eye-watering one hundred degrees proof, or 58 per cent ABV (alcohol by volume) – but as the hydrometer was yet to be invented, alcoholic strength had to be checked another way. If there was one thing a warship had in abundance it was gunpowder and it was established that this would fail to ignite if soaked in rum that contained too much water – anything less than 57.15 per cent ABV. It is perhaps worth mentioning here that while 'proof' today stands synonymous with 'evidence' or 'validation', its original meaning was more akin to 'test', as in 'the proof of the pudding is in the eating'.

This was the standard test conducted by the captain of any ship to make sure he was not being diddled by the supplier. The crew would also conduct spot checks throughout the voyage to make sure the purser was not watering down their beloved rum to make a bob or two on the side.

Although such fun-time test methods were banned in the UK in 1980 by Brussels officials who ruled that all alcohol must be stated by ABV, some imported brands of vodka and rum from Poland and the Caribbean still state the strength of the contents by the old proof-ratings: this makes sense of labels proclaiming, say, 160° proof. For the adventurous and possibly suicidal drinker, the strongest legal drink on the market is Spirytus Delikatesowy, a Polish vodka with a 192° proof rating – just short of pure alcohol.

HOGSHEADS AND FIRKINS

The size of the old wooden barrels used to transport wine and beer was in the main dictated by the size of the doorways of the inns and taverns to which they would be delivered. Although in general speech 'barrel' is used of any wooden cask, it is in fact a specific size of one containing thirty-six UK gallons (31.5 US gal.).

In France, the standard wine barrel, called a *barrique*, contained 225 l / 396 pt and, like the English beer barrel, had to be of a size that rendered it manoeuvrable through the narrow streets and tavern doorways of old Paris. This resulted in the *barrique* standing disproportionally high so that, when filled with cobblestones dug out of those same streets, six of them abreast made a perfect 'barricade' to block troops' advances during times of riot and unrest.

Irked by the speed at which vast swathes of the city could be rendered no-go areas to his troops by wine barrels, in 1853 Napoleon III (1808–73) commissioned Baron Haussmann to demolish most of medieval Paris and impose the now much-loved network of avenues and boulevards which, too wide to barricade, favoured the deployment of massed troops and cavalry charges. Thus did the dimensions of the humble wine barrel contribute to the beauty of Paris we see today.

OIL BARRELS AND STEEL BANDS

In 1483, Richard III (1452–85) decreed that the butts (or casks) used for the wholesale of wine should be none other than the seventy-gallon puncheon and the thirty-five-gallon tierce, this second measure surviving to travel to the New World where it equated to forty-two of the slightly smaller US gallons.

Centuries later, in 1858, Edwin Drake (1819–80) became the first man to actually drill for oil instead of scooping it out of natural seepage pits. Ridiculed for pursuing a madcap venture, Drake, surrounded by jeering locals at Titusville, Pennsylvania, made the world's first oil strike, collecting his black gold in old washtubs bought for the purpose.

Distressed by spillage levels during transport and reasoning that a washtub was but a barrel without a lid, Drake turned to the local whisky makers who were still using King Richard's forty-two-US-gallon tierce. But his levels of production gradually forced down the price of oil from $10 a barrel at the time of his strike ($330 at today's values) to $2 a barrel by 1860 and, as that was about the same as the cost of a new barrel to ship his oil, bulk rail transport evolved but shipments were still quoted in multiples of the forty-two-US-gallon barrel.

The US Military of the twentieth century preferred to store its oil in a steel fifty-five-gallon (forty-six-imperial-gallon) steel drum, made especially for them by the Iron Clad Manufacturing Company, owned by celebrated journalist Nellie Bly (1864–1922). It was abandoned examples of such containers, left lying around the island of Trinidad by the US Navy of the 1940s, that locals adapted to musical use.

It was thus no coincidence that the first properly sponsored and organized troupe was the Esso Trinidad Steel Band, which went on a world tour in 1967 to promote the eponymous oil company.

WEIGHT, DISPLACEMENT AND DENSITY

As outlined below, the concept of weight is a fiction we have invented to explain the effects of gravity on ourselves and anything else that has mass: in space, there is no gravitational pull, and so no weight. But mass is only one such consideration. In any matter, the closer the atoms are packed together, the greater the density. Osmium, twice the weight of lead for any given volume, is the densest element on the planet. Within a black hole, it is conjectured that matter becomes so dense that a mere handful would outweigh the earth, resulting in a gravitational pull of such strength that even light cannot escape it. This, and the fact that light bends when passing a planet, raises for some the question as to whether light itself has mass. Mainstream opinion maintains that light has no mass, only energy and momentum, but scientific opinion is so often forced to execute a U-turn that this may yet change. *Eppur si muove*, and all that!

THE SOUL

The rather convoluted movie *21 Grams* (2003) was inspired by the research of Dr Duncan MacDougall (1866–1920) of Haverhill, Massachusetts, who, in 1901, tried to establish the weight of the human soul.

Having constructed a platform incorporating a beam balance sensitive to two-tenths of an ounce (5.6 g), MacDougall selected six terminal patients from a local hospice and, one by one, had their beds transferred to the platform, as their deaths were judged imminent. The first patient, a man dying of tuberculosis, spent nearly four hours on the platform before dying and, according to MacDougall, at the point of death the balance beam dropped to hit the stop bar with an audible bang. The sudden weight loss was three-quarters of an ounce, or 21.3 g.

Two other patients gave similar weight-loss readings, but the other half of the inadequately large test group were discounted: one died while MacDougall was tinkering with the settings on the beam; another had a nurse bend over him at the point of death, no doubt exerting some pressure on the bed; and the penultimate case gave erratic readings. MacDougall then repeated the experiment with fifteen euthanized dogs, which gave no weight variance at the point of death, prompting him to assume that this was because dogs have no soul: obviously a man who had never enjoyed their company.

Some conjectured that the weight loss claimed by MacDougall might have been attributable to the final exhalation of the dying, but the lungs simply cannot hold that much air. Similarly, the evaporation of sweat would not be so acute, and any other kind of bodily evacuation at the point of death would

DORD

One of the most famous errors to be promoted by any dictionary of merit, this ghost word is still held by some to be a synonym for density.

In the 1934 edition of *The New International Dictionary* by Merriam (now Merriam-Webster), the relevant entry should have been headed 'Density or D. or d.', to indicate that in the ensuing text the letter d, be it capped up or not, would stand for 'density' to save space. Unfortunately, things got a bit scrambled in the editorial process, resulting in the heading going to print as 'Density or Dord'.

Failing to spot their own error, compilers of that same edition gave 'dord' its own entry on page 771, where readers were advised that the term was a noun from the field of 'Physics & Chemistry' before advising them to see the entry under 'Density'. The error went unchallenged for years resulting in its inclusion in other dictionaries and with some today still quoting the dord-factor of various materials.

remain on the bed or platform: it was just a badly conducted study with an inadequately sized sample group. Either way, the hostility of the debate following his going public with his rather shaky findings dissuaded MacDougall – and anyone else, for that matter – from conducting similar experiments, so twenty-one grams remains the much-quoted weight of the human soul.

EUREKA!

The most famous story attached to the problems of measuring displacement and density is that of ancient Greek mathematician Archimedes (c. 287–c. 212 BC) getting into his bath to ponder his latest problem. Hiero II, King of Syracuse (c. 308–215 BC), had commissioned a crown of pure gold from a local craftsman, whom the tyrant suspected of having adulterated the gold with other base metals so he could pocket the difference. Hiero demanded that Archimedes prove the case, one way or the other.

Archimedes knew how much gold had been provided and that such an amount of gold would have a specific volume, but how was he supposed to check that volume now that said gold had been worked into an irregular shape? As he stepped into his bath he noted how the water level rose and realized he had the answer; immerse the crown in water and measure the volume of the water it displaced. According to legend, the poor chap lost his head in all the excitement, prompting him to run naked through the city streets shouting 'Eureka!', or 'I have found it!'.

Actually, the Greek term is really *heureke*, pronounced 'ev-reeka', but let's not get pedantic. Hiero's crown was subsequently tested in a small bowl and proved adulterated, so the goldsmith lost his head too – but in a far more literal manner.

BMI

With more people dying today in the developed world from obesity-related conditions than die from starvation in the developing world, we should all perhaps watch our waistlines. But calculating your BMI, or body mass index, at the local clinic or gym is a complete waste of time and money.

The usual method divides the number of kilograms you weigh by your height squared in metres. Simple… and totally useless, as two people of the same height and weight could have completely different body shapes and be carrying fat in different places about their bodies.

The only way to establish a meaningful BMI is to go back to Archimedes in his bath. By opting for a hydrostatic test, you are lowered into a small pool of water on a cradle of known mass. With fat having a dramatically different density, and therefore volume per kilogram, than muscle tissue, only this will give you a figure of any accuracy and meaning.

SHIPS AND THEIR WEIGHT

Few landlubbers understand the weights given for ships, as none involves the vessel being put on a massive set of scales.

One way to estimate a ship's tonnage again harks back to Archimedes and his principles of displacement. With the ship fully crewed, provisioned and fuelled, and with all cargo loaded, her displacement tonnage is calculated by the volume of water she displaces, with every thirty-five cubic feet (1 m³) of water equal to one displacement ton.

Next there is deadweight tonnage, which notes the difference between the ships' displacement tonnage while she rides at anchor, fully provisioned and fuelled, this being deducted from her displacement tonnage after all cargo is on board.

A ship's gross registered tonnage is worked out by calculating all her enclosed space and then allowing one ton per hundred cubic feet (3 m³). In practice, merchant ships are generally quoted by their gross or deadweight tonnage, with displacement tonnage usually reserved for warships.

PLIMSOLL LINES

The white lines on the fronts of ships' hulls are there to ensure the vessels are not running on too great a displacement tonnage for their own safety, as even the casual observer can see how low they are riding in the water.

First championed by the British MP Samuel Plimsoll (1824–98) in the late 1850s, but not made compulsory until the Merchant Shipping Act of 1876, these lines have to take into account the fact that fresh water and sea water have different densities, with the former rated at 62.4 lb/ft^3 / 999.6 kg/m^3 and the latter 64 lb/ft^3 / 1,025 kg/m^3. As ambient temperature can cause significant and dangerous variance in those figures as the ship rides lower the less dense the water, the lines are scaled as follows:

TF – Tropical Fresh Water

F – Fresh Water

TS – Tropical Seawater

S – Summer Temperate Seawater

W – Winter Temperate Seawater

WNA – Winter North Atlantic

Although first made in the 1830s, the light shoe originally made from bleached canvas with a white rubber strip joining the upper to the sole, was nicknamed a plimsoll in the late 1870s for its similarity in appearance to Samuel Plimsoll's lines on ships' hulls. The incorrect variant 'plimsole' arose through confusion with 'sole'.

SPECIFIC GRAVITY

An oddity in the world of measurement, specific gravity is a scale with no units, as it relates to pure water at its maximum density (at an ambient temperature of 3.98 °C / 39.16 °F). A hydrometer, a sealed glass float with a calibrated stem, is used to ascertain the specific gravity, or relative density, of a fluid: a specific gravity of four means that a fluid is four times as dense as pure water.

The corresponding density of air is expressed as being 1.29 kg/m³ / 0.08 lb/ft³, which, in conjunction with the pressure of one atmosphere at sea level, explains why you can drink through a straw. But don't imagine for one minute that you are sucking the fluid up the tube – it is atmospheric pressure that pushes the liquid up the straw and into your mouth. When you suck on a straw, you remove the air from the inside, leaving the air pressure outside pressing down on the fluid in the glass with one atmosphere at sea level – 14.7 lb/in.²/ 1 kgf/cm² – which forces the drink up the straw. That is about the weight of one sixteen-pound ten-pin bowling ball per square inch (1 kgf/cm²), or when drinking out of a glass of 2.5 in. / 6.35 cm diameter, about five such bowling balls: a suction strength no human could generate with their mouth alone.

Even with this vacuum, the pressure of one atmosphere is incapable of pushing up a column of water – or cola – any higher than about thirty feet or nine metres (depending on how high you are above sea level). To lift it any higher you would have to generate pressure from below, rather than suction from above.

Obviously, fluids of greater relative gravity have their column height further restricted due to increased weight, which is why mercury, with a maximum column lift of about 2.5 ft / 76.2 cm, was chosen for barometers. These restrictions hold true no matter the internal diameter of the straw, tube or pipe. Of course, no one can drink through a straw of any length in space, as there is no atmosphere to do the hard work for you.

GRAVITY AND WEIGHT

As stated at the beginning of this chapter, there is, strictly speaking, no such thing as weight, as it is simply a man-made concept. Weight was invented to make sense of the effect of the earth's gravity field pulling objects back to the planet's surface: in space nothing has weight.

Equally, and allowing for the moment that you could withstand the heat and pressure, were you able to stand at the centre of the earth, you would have no weight at all, as in space. Just as time incrementally slows the further you travel from earth (see page 119), the higher up you go, the less the gravitational pull on your body trying to return it to the earth's surface.

DON'T TRY THIS AT HOME!

In any specific environment gravity is a constant, affecting all bodies in motion or those at rest. Many will be familiar with the apocryphal yarn of Galileo Galilei (1564–1642) dropping a cannonball and a musket ball from the Leaning Tower of Pisa to demonstrate that they both hit the ground simultaneously, despite the disparity in weight. Few extrapolate that the same will hold true in a far more dangerous ammunitions demonstration.

Hold a pistol in your right hand and a spare bullet in your left. Extend the right arm to the firing position and hold the spare bullet level with the barrel. Fire the gun horizontally and release the spare round at the same time and, as did the Pisa cannonball and bullet, both your bullets will hit the ground simultaneously. This will hold true whether you are standing on the ground or firing the gun from the top of a hundred-foot tower. No matter that it travels a greater distance and at a much higher speed than the dropped round, the bullet fired is subject to exactly the same gravitational force pulling it down to the ground. In a slightly less dramatic demonstration of the same effect, no matter how fast you are walking or running when you drop your keys, they will always hit the ground beside you. Now, that you *can* try at home.

Should a person 'weighing' 150 lb / 68 kg climb 10,000 ft / 3,048 m up a mountain their weight would drop to 149.92 lb. Similarly, and because the earth bulges significantly at the equator, that same person would tip the scales at 150 lb at the North Pole but only register 149.75 lb at the equator. We all have mass and density – but no weight.

Thus it is clear that no set of scales, from the hand-held ones you use to weigh your luggage before setting off for the airport to the tare bridges used to weigh lorries, measures weight but, rather, gravitational pull – they just express the result in figures that make sense to us.

THE GRAM

First mention of the gram in Europe is noted in the anonymous 'Carmen de Ponderibus et Mensuris', or 'Poem about Weights and Measures', a jaunty little ditty published *c*. AD 400. The earliest French use of the term came in medieval times to denote a weight based on the ancient Greek *obol*. This was a thin spit of copper or bronze, about the length of a small nail, which served as a unit of both currency and weight, and in ancient Athens one *obol* would have bought you three litres or five pints of wine.

Either way, a handful of *oboli* was a recognized unit of money in business circles and, with the Greek for a handful being *drakhme*, survived for centuries as the *drachma*.

On 7 April 1795, with France midway through the upheaval of its revolution, the gram was redefined as being equal to the weight of one cubic centimetre of pure water weighed at its maximum density in a vacuum. Water is at its greatest density at the point when it just melts from ice, which in 1795 was thought to be at the temperature of four degrees Celsius (39.2 °F); it is now known that water hits its maximum density at 3.984 °C / 39.17 °F.

With the kilogram arrived at by the simple expedient of multiplying the gram by one thousand, this fractional error in temperature resulted in the prototype kilogram block, made from an alloy of 90 per cent platinum and 10 per cent iridium, being fractionally short by twenty-five parts per million or, in simple terms, about the weight of one grain of dry rice. But wrong is wrong, and as more sophisticated measuring and weighing equipment became available, along with knowledge of the true temperature at which water attains its maximum density, the error was rectified.

THE OUNCE

The modern ounce is somewhat lighter than its progenitor, the Roman *uncial*, which constituted one twelfth of the Roman pound, whereas today there are, of course, sixteen ounces to the modern pound. For more on that old Roman pound see TROY WEIGHT (page 87).

Not until the International Conference on Yards and Pounds of 1959 did the ounce achieve stability at 28.3495 g, with the pound likewise agreed to be 0.45359237 kg. But the ounce's currency in the UK would be short-lived: although still a functioning unit in the USA, it was outlawed as a vending unit in the UK by an EU directive in 2000.

Prior to the ounce's standardization it had many definitions. The Troy and the Apothecary's ounce equated to 31.1 g while most European ounces ranged from about twenty-seven to thirty grams, the Dutch one-hundred-gram ounce being a notable exception. But most popular throughout eighteenth-century Europe was the Maria Theresa ounce, as that was tied to the exact weight of a silver coin then in wide circulation.

Queen of Austria-Hungary from 1740 to 80, Maria Theresa's (1717–80) one-ounce thalers were internationally accepted as a twenty-eight-grams weight of pure silver. The coin's name means valley, as they had first been struck from silver lodes discovered in Joachimsthal, or Jacob's Valley, in Bohemia, which she also ruled. In time, these coins travelled with the immigrants to the New World, where the name altered from thaler to dollar.

THE POUND

This is a direct descendant of the Roman *libra pondo*, one pound by weight, which seems to have varied between 322 and 329 g. The term 'pound' was introduced to England by the fifth-century invasion of the Saxons, who seem to have confused their *libras* with their *pondos*; it was *libra* (meaning scales) that specified the unit, which is why the abbreviation for pound is still lb.

This early twelve-ounce pound was defined as being equal to the weight of 5,760 grains of barley, which would now be on par with 373.26 g, but for bulkier goods the sixteen-ounce pound of 7,000 grains (453 g) is noted as early as the thirteenth century. By the mid-fourteenth century most of the English wool production was being exported to the weavers and dyers of Florence, who worked on a pound based on the Florentine ounce of 437 grains so, to encourage this trade, Edward III (1312–77) adopted this unit to generate a new sixteen-ounce pound equal to 6,992 grains.

That said, for two hundred years after Edward III's reign various different pounds were used throughout England, depending on the goods being weighed. Expensive wares such as butter, cheese and meat tended to be sold by the old Roman twelve-ounce pound while low-value goods such as hay, malt and hops were sold by the obsolete twenty-seven-ounce pound.

Until the sixteen-ounce pound went the way of the ounce in 2000, this and divisions thereof were the only legal weight for retailed goods in the UK – especially bread (see below). The Assizes of Bread and Ale in 1266 declared it illegal to sell anything other than a one-pound loaf and the penalties for infringement were so severe that, on orders for the local manor and any other place likely to have scales, bakers always threw in one free loaf per dozen, just to be on the safe side – hence thirteen items being commonly described as a baker's dozen.

THE POUND LOAF

Reinforced by the Bread Act of 1822, the one-pound loaf was reduced to 0.88 lb or fourteen oz (396.9 g) during the Second World War to eke out food-reserves. This remained in place until the UK went metric, when it was argued that it would make more sense to round it up to 400 g, and to also make available an 800 g (1.7 lb) loaf. The changeover of baking racks and tins was implemented on the May Bank Holiday of 1977, after which it was illegal to even make a one-pound loaf, let alone sell one.

This remained the case until 2008 when an EU directive ruled it discriminatory to UK bakers to restrict their sales weights while their European counterparts could produce breads of any weight they fancied.

POUND STERLING

Although it underwent decimalization in 1971, the pound sterling is the oldest currency in circulation. Its roots lie in the silver pennies that started to circulate in eighth-century England, when two hundred and forty of them could be struck from the long-redundant Tower pound, which equated to 350 g instead of the formal pound of 453.5 g. This remained the sterling standard until 1526 when Henry VIII (1491–1547) decreed that the Tower pound would be replaced for minting purposes by the 373-g Troyes pound.

'Sterling' is a corruption of starling (from the Norman French *esterlin*), as each thin penny had a star in the form of an indented cross on the back to allow it to be cut up for small change – halfpennies and farthings. Early coins held their worth by virtue of the silver they comprised, and this was common to many coins such as the Spanish silver peso, otherwise known as the piece of eight, which circulated in the New World before the dollar, and is why North Americans still talk in terms of 'two bits' or 'four bits'.

Subsequent monarchs, such as the aforementioned and avaricious Henry VIII,

King Henry VIII

debased the pound sterling. Henry adulterated each of his newly regulated Troyes pounds of silver with base metals to get seven hundred and twenty pennies to the pound of silver employed. Such coins are still legal tender in the UK but are only struck for the traditional royal handout of Maundy money on the day before Good Friday. This ancient tradition is governed by the age of the monarch at the time, so in this year of writing, with Elizabeth II (b. 1926) being ninety, the ninety men and ninety women selected will each receive ninety silver pence at the cathedral chosen to be the venue.

The £ sign is basically a stylized 'L', as in the Latin *libra*, while the $ sign is reminiscent of the old style '8', which resembled a capital 'S' with a slash through it, to indicate that the new dollar, made official US currency in 1792, held the same value as its predecessor, the Spanish piece of eight.

TROY WEIGHT

As early as the opening of the ninth century, traders from across Europe were making regular visits to the lucrative markets at Troyes, a commercial centre for everything from spices to precious metals and gems that lay southeast of Paris. But visitors had to keep their wits about them, as the town ran on its own rather unique system of weights and measures that embraced the Troyes pound, the Troyes ounce, the Troyes pennyweight and the Troyes grain.

Tipping the scales at 31.1 g, equal to 1.097 ounces avoirdupois, the Troyes, or troy, ounce was about 10 per cent larger than most other ounces around at the time, so it was a common error to assume that the troy pound would likewise be larger. Modelled in part on the old Roman pound and ounce, the troy pound in fact only had twelve ounces to its pound, or 5,760 troy grains, while the pound avoirdupois hefted seven thousand troy grains, or 373.25 g instead of 453.59 g.

PENNYWEIGHTS AND SCRUPLES

The twenty-four-grain pennyweight is long obsolete, but the grain persists in modern small-arms ammunition to define the precise weight of the bullet, or the number of grains of propellant required to expel it: the notorious .357 Magnum, for example, being a load of a hundred and fifty-eight grains. Until the twentieth century, the grain also featured in medicinal preparations and recommended dosage, along with its larger relative the scruple, equal to twenty grains.

Common to many European languages, 'scruple' derives from the Latin *scrupulus*, a small, sharp stone, as used in conjunction with the *calculus*, or larger and rounded pebble, in the ancient Roman abacus (hence 'calculate'). Rarely seen outside the world of fringe medicine after the 1920s, both the grain and the scruple were banned from any further pharmaceutical use throughout the EEC in 1971.

GOLD AND DIAMONDS

As for the troy weights system itself, this is alive and well. Dealers in the gems and precious metals business still use troy ounces and pounds, the old troy grain and the carat. This latter unit is an Eastern reflection of the old European use of various cereal grains as a standard unit of measure, and one that relates to the extremely light but consistent weight of the carob seed, known in Arabic as the *qirat*.

Until an international agreement of 1907 the carat varied from country to country, with Italy unable to agree within its own borders; in Florence it was 197.2 mg, but 207 mg in Venice. Most other countries worked to fractions above or below 205 mg but the agreement of 1907 pegged the carat to a neat 200 mg, this overriding a previous agreement of 1895 pegging the carat to a rather clumsy 205.4152 mg.

One of the largest diamonds ever found was the Cullinan, discovered in South Africa in 1905, with a rough weight of 3,106.75 carats; to put that into perspective, there are 1,866 carats to the troy pound and 2,268 to the pound avoirdupois.

The carat is a pretty straightforward weight, but its use in the definition of the fineness of gold is a bit more complicated. The troy ounce was divisible into 24-ounce carats, so the carat was likewise perceived as being one twenty-fourth part of any whole: no matter the weight of an item, the

The Cullinan Diamond

gold contents are perceived in multiple twenty-fourths, with 24 carat indicating pure gold.

THE BUSHEL

To stand as intermediary measures between the pound and the ton, the Normans ushered into England the peck and the bushel. The peck was a unit of dry volume, equal to the weight of sixteen imperial pints of water (9 l), which was inspired by the old Roman *modius*, used to issue grain rations to legionaries in units of 8.7 litres. The bushel, with its name likely derived from the Old French *boissel*, a small box, contained four pecks.

In Norman England, where the litre was yet to appear, the peck was rounded to sixteen pints, or two gallons, with the bushel standing at eight gallons. Come the sixteenth century, when the avoirdupois system was gaining ground over other English systems, the bushel was redefined to fifty-six pounds and the peck to fourteen pounds. The peck now largely belongs to history while the bushel is still functional, but no longer a fixed unit.

In the UK the bushel survived as a rarely used measure equating to eight imperial gallons until it was excluded from trade by an Act of 1985, but it is still much used in the USA as a measure of grain and cereal. An unusual measure, it varies in weight depending on the moisture content of the commodity. In the USA a bushel of oats, for example, weighs thirty-two pounds (15 kg), but a bushel of soybeans weighs in at sixty pounds (27 kg).

From the time of its introduction by the Normans, 'bushel' could also describe the wooden box of fixed dimensions (2,219.3 in^3/ 36,367.8 cm^3) used to calculate it.

TALENTS AND SHEKELS

Deriving its name from the ancient Hebrew for weighing, the shekel started out as a small measure of grain equivalent to about eleven grams (0.4 oz) and, when later taken as a weight for the silver coin of the same name, tended to yo-yo between nine grams and seventeen grams (0.3 and 0.6 oz), depending on the caprice of the issuing authority of Judea.

Next came the mina, a unit of sixty shekels, and finally the talent, of sixty minas. Originally a Greek unit, the talent was the mass of water required to fill an amphora, each country using a different size. The Athenian amphoral talent stood at a mass-weight of twenty-six kilograms or fifty-seven pounds, whereas the Roman weighed just over thirty-two kilograms or seventy-one pounds.

THE HOBBIT

This was an extremely confusing and varying weight favoured in rural Wales until it was banned throughout the UK in the mid-to-late nineteenth century for all the trouble and convoluted litigation it engendered.

A Welsh hobbit of oats weighed 105 lb (48 kg) but for barley and wheat it was 147 lb (67 kg) and 168 lb (76 kg) respectively. In Flintshire alone, a hobbit of potatoes weighed 200 lb (91 kg) but this rose to 210 lb (95 kg) if the commodity being measured was new potatoes.

With English traders forever taking delivery of a smaller quantity than anticipated, connotations of shortness and cunning crept in. The Exchequer of Pleas endured the litigation of *Tyson v. Thomas* (1825) and *Hughes v. Humphreys* (1852), both of which reduced the judges to tantrums of frustration as all involved argued the traditional and regional weight of a hobbit. This judicial frustration attracted much lampooning of the hobbit in various publications, including Charles Dickens' weekly magazine *All the Year Round* (1859–95).

J. R. R. Tolkien (1892–1973) loved Wales and the Welsh language. He spoke both medieval and modern Welsh and lectured on the subjects at Leeds University. Although he always claimed to have invented 'hobbit' as the name for the short-statured protagonists in his famous novels, it is likely that he was familiar with the unit that had caused so much argument over short-weighing in the previous century.

PLAYING THE TABLES

The periodic table presents the pattern that it does because the man who devised it, a rather eccentric nineteenth-century chemist and professor at the University of St Petersburg named Dmitri Mendeleev (1834–1907), was an avid card player.

Mendeleev is said to have enjoyed arranging cards on which he had written the names of all the chemical elements before organizing them into suits according to their properties. His passion for the card game known as solitaire, or patience, is said to have bordered on obsession, and it was in 1869, when he laid out his periodic 'suits' in columns, as in the game of solitaire, and ordered them by the ascending atomic weight of the elements inscribed, that the format of the periodic table became clear. The table's orderly structure enabled Mendeleev to predict the atomic mass of then-undiscovered elements that belonged in its gaps and sequences.

DATES AND
CALENDARS

With humans appearing to be the only animals on the planet with an understanding of the ageing process and their own mortality, it should come as no surprise that the measurement of time has been important to us ever since we got past the grunting stage.

Most early religions were matriarchal and lunar-orientated, so the lunar cycle was taken as the prime measure of the passing of time as reflected by the two oldest time records, both of which were found in Africa. The twenty-thousand-year-old Ishango bone, unearthed in the Congo in 1960, seems to mark off the days of six lunar months. Claudia Zaslavsky, author of the intriguing book *Africa Counts* (1973), believes it to have been a woman's counter for her menstrual cycle. With age estimates ranging from thirty-five thousand to forty-four thousand years, the much older Lebombo Bone, discovered in the mountains between South Africa and Swaziland, is inscribed with twenty-nine notches, the number of days in a lunar month.

Ancient calendars evolved from the need to know the right days for religious observance and the payment of debts or rents. As ancient Rome ran on the old-style lunar calendar, the city's priests would stand vigil to announce their sighting of the new moon, so the first religious calendars took their name from the Latin *calare*, to call or shout out.

The first day of each Roman month was thus known as the *calends*, which gave rise to 'calendar': first a term for an accounts book. It was much the same in ancient Persia, where a ledger was called a *devan*, with the debtor sitting on a specially shaped divan to receive payments. Medieval England also ran on lunar calendars, which were hand-painted by monks who highlighted important festivals and rent-owing days in red lead, hence 'red-letter days'.

BY THE RIVERS OF...

The ancient city of Babylon was an international trading centre long before the rise of Greco-Roman culture. The city-state worked on the old lunar calendar of twelve months, alternating between twenty-nine and thirty days to allow for the fact that there are in fact 29.53 days to the lunar cycle. The old Babylonian year of 348 days (instead of the more accurate lunar year of 354.36 days) presented them with complex problems when they later shifted to the lunar-solar calendar, with the solar year comprising 365.24 days.

But we owe it to the Babylonians for there being seven days in a week and for the structure of hours and minutes. Astrology was of prime importance to the Babylonians, who knew of only seven significant celestial bodies – the Moon, the Sun, Mars, Mercury, Jupiter, Saturn and Venus – and invented the seven-day week to allocate a day of veneration to each. The Babylonians and the Egyptians seem to have come up with the zodiac between them, after noting how the constellations seemed to move about the sky with the seasons, but lost in the mists of time are the identities of those who decided to join up the dots to give us the star signs we still recognize today.

THE ZODIAC

For those who do believe in astrology there are a couple of problems worth taking into account, the first of which is the omitted thirteenth sign of the zodiac.

The Babylonians were either unaware of the thirteenth house or chose to studiously ignore it, as it made a nonsense of their numerical obsessions. Either way, other more accurate astrologers since the second century BC have taken into account the sign of Ophiuchus, or the serpent bearer, which rises to solar prominence between 29 November and 19 December to create an astrological sign to sit between Scorpio (more properly Scorpius), and Sagittarius.

Until Julius Caesar (100–44 BC) reformed the calendar from a ten-month year to a twelve-month one, the Roman zodiac comprised only eleven signs: those born under the influence of Libra have Caesar to thank for their sign, which had to be hurriedly invented and shoe-horned into the zodiac to sit within the new structure of the year. On top of that, none of the signs today are where they were for the Romans, the Greeks or the Babylonians.

Thanks to over two thousand years of the gravitational pull inflicted on the earth by the moon and the sun, the earth's position relative to the constellations has changed, meaning the signs of the zodiac have been gradually shunted back to now stand about a month out of sync with modern astrological charts. Capricorn, for example, now rises between 20 January and 16 February, while astrological charts still mark it from 22 December to 19 January. Capricorns are therefore actually Aquarians, and so forth back through the charts.

A BURNING ISSUE

In 1930 the *Sunday Express* became the world's first newspaper to run an astrological column, with both the editor, John Gordon, and the star-caster, Richard Naylor, technically risking their lives.

On 24 August, Gordon published Naylor's chart predicting the life of the newly born Princess Margaret (1930–2002), a crime of treason for which both could have been legally executed. Whether or not you believe that lumps of rock millions of miles out in space can orchestrate your life is a matter of personal choice but, since the late fourteenth century, it has been deemed treason in the UK to predict by astrology the life of the monarch or any of his or her immediate family – just in case the 'science' were accurate.

If you could foretell the day of the monarch's death you could stand ready for a coup at precisely the right time and, in 1930, the death penalty was still very much in place for treason. The casting of stars for members of the British Royal Family is still technically illegal but, in practice, not considered worth the ridicule of prosecution.

FROM JULIAN TO GREGORIAN

The old Roman ten-month year began in March, as seen in the names of September, October, November and December, so named from the Latin for seven, eight, nine and ten. When the additional months of January and February were added in the Julian calendar, these months slipped forward in the year to stand, as they do today, as the ninth, tenth, eleventh and twelfth months of the year. But the year for the Romans still turned over in March, with February the twelfth and final month of the year.

Instituted in 46 BC and named after Julius Caesar, the Julian calendar was a better system, but still flawed. Caesar's general Mark Antony (83–30 BC) immediately suggested that the seventh month be named in honour of Julius, who promptly stole a day from February to increase his eponymous month from thirty days to thirty-one. Later, Augustus Caesar (63 BC –AD 14) felt slighted that August, the month named after him, only had thirty days, so he too stole another from February to put him on par with Julius and leave poor old February two days short.

The main problem with the Julian calendar was that it generated too many leap years. The earth takes 365 days, five hours, forty-

Julius Caesar

eight minutes and forty-five seconds to orbit the sun, so the Julian calendar added an extra day to February every four years. But this was an over-compensation which, come the sixteenth century, left the calendar twenty-four days out of sync with the equinoxes and solstices

THE GREGORIAN CALENDAR

The usurping Gregorian calendar was devised by Italian astronomer Luigi Lilio (*c.* 1510–76) who unfortunately died before he had convinced the Vatican to adopt his system. The baton passed to the hands of the German-Jesuit mathematician Christopher Clavius (1538–1612) who oversaw its institution under the aegis of Pope Gregory XIII (1502–85), hence the name. Unlike the previous Julian calendar, this only allowed for leap years if the date was evenly divisible by both four or, if it was a centurial year, divisible by both four and one hundred. Thus centurial years are only leap years if evenly divisible by four hundred, so 1700, 1800 and 1900 were not leap years, but 2000 was. Also, by incorporating only ninety-seven leap years every four centuries, the Gregorian calendar stays in step with the solar or tropical year, which is only out of sync by one day every 3,300 years.

Announced by papal bull in 1582, Gregory decreed that ten days had to be lost to facilitate the changeover so in that year of decree the faithful would go to bed on 4 October and wake up on 15 October. Most Catholic countries adopted the new system straightaway but, suspicious of it all being some dastardly popish plot, most Protestant countries shunned Gregory's newfangled calendar. This proved to be a doomed strategy, as one half of the world became increasingly out of step with the other.

Germany, the Netherlands, Switzerland and most of Scandinavia capitulated in 1700, and the UK grudgingly followed, along with its colonies, in 1752, but Russia refused to change until 1918, Greece until 1923, and Turkey finally fell into line in 1927. Only the Berbers of North Africa and the religious, all-male enclaves of Mount Athos in Greece still run on the Julian calendar, or modified versions thereof.

Sweden simply didn't know what to do. Thinking it best to opt for gradual change, they reasoned that if they resisted the temptation to observe Julian leap years from 1700 through to 1740 they would find themselves in step with the Gregorian calendar. But somehow the plan was forgotten as Sweden happily went leap in 1704 and again in 1708, to leave itself out of step with everyone else on the planet. Throwing up their hands in exasperation, the Swedes announced they were going back to their old Julian ways with a double leap year in 1712, and a thirty-day February. Finally, in 1753, they threw out eleven days and went Gregorian.

Naturally, the longer a country delayed the change, the more days had to be abandoned; the UK lost eleven days in 1752 with late starters Turkey, Russia and Greece having to lose thirteen days. But it was in Britain that there was the most confusion, with people mistakenly believing they had been cheated out of eleven days' pay and would die eleven days before their time. It is widely stated that they took to the streets, raising the cry 'Give us back our eleven days!', but this is a Georgian urban myth started by the painter and engraver William Hogarth (1697–1764).

THE RITES AND WRONGS OF SPRING

Another possible ramification of the shift to the Gregorian calendar is the tradition of April Fools' Day. Because the old New Year festivities culminated on 1 April, even after the calendar change people continued to mark the old festival. Indeed, we are talking about the 1750s, a time when news travelled slowly, so it is more than likely that folks in rural districts remained blissfully unaware of the new calendar for quite some time, leaving them celebrating the 'wrong' day. This, no doubt, would have been taken as an indication of stupidity by their more clued-up urban cousins.

While it is true that there are other, much older festivals of upheaval and chaos to mark the commencement of spring, it was not until the adoption of the Gregorian calendar that playing pranks on the unwary became common in the UK so, if not the progenitor of the tradition per se, the institution of the Gregorian calendar can certainly claim a considerable responsibility for April Fools' Day.

THE MISSING DAYS

Few in the UK were pleased by the imposition of what was deemed a popish calendar, and it became one of the debating points between the Whigs and Tories in the 1754 election. William Hogarth's *An Election Entertainment* (1755) shows a gathering of drunken Whigs and camp-followers in a tavern with a stolen Tory banner lying on the floor in the foreground, inscribed with a demand for the return of the eleven days.

Although there were no actual riots, confusion mounted with the approach of the tax day of 25 March 1753. Prior to

An Election Entertainment *by William Hogarth*

the British adoption of the Gregorian calendar the year began, as did the old Roman or Julian year, in March, with a festival commencing on 25 March, Lady Day, leading up to its octave on 1 April – a perfectly sensible seasonal opening of the new year. Because of the abandonment of eleven days the banking and business fraternity demanded that their day of reckoning with the taxman be likewise shunted forward to 5 April, which explains why the UK tax year still turns on that date today.

MAYAN CALENDAR

Properly known as the Mesoamerican Long Count calendar, this was an incredibly intricate piece of stonework that detailed the passing of about 5,125 years, from 3114 BC to the day we call 21 December 2012. Many will remember the headlines circulating in the summer of 2011, warning the gullible that the Maya of Central America had foreseen the end of the world, as 'proven' by the fact that their calendar stopped on the aforementioned day of doom and destruction.

The first problem with the perception of the calendar as an indication of Mayan prophetic ability is the fact that they did not devise the system but inherited it from the preceding Olmec culture. The second problem is the fact that the Long Count calendar is wildly inaccurate as, although solar-based, it did not allow for leap years, so, going one day awry every four years, it did not even end on the day we call 21 December 2012.

But the third problem and the final nail in the prophecy coffin is the fact that the calendar worked in bak'tuns, or blocks of 400 years, and at the end of the thirteenth bak'tun – allegedly 21 December 2012 – readers of the calendar were expected to go back to the first bak'tun and start all over again.

Either way, we are still here. As for the prophetic capabilities of the Maya, they didn't even foresee their own tragic destruction at the hands of the Spanish in the sixteenth century – had they done so they would have stopped chiselling when they got to 1546, not 2012.

THE ISLAMIC CALENDAR

Even before the institution of Islam, the Arab world ran on a lunar calendar, which is why the symbol of the crescent moon has long been of great importance in those climes. In the year the West now calls AD 634, the Muslim Caliph Umar (c. 583–644) set his scholars to work to refine the old lunar calendar by relating it to the life of the prophet Muhammad (d. AD 632) and establishing Year 1, corresponding to the present Western year of AD 622, as the year in which Muhammad fled Mecca for the comparative safety of Medina. Because of this, all years in the Islamic calendar carry the designation AH which stands for the Latin *Anno Hegirae*, or Year of the Hegira (his journey).

Due to the vagaries of the lunar cycles that disjoint the lunar year from the solar, all Islamic festivals shift about from year to year, with New Year's Day of our year 2008 (1430 AH) falling on the day we call 29 December; in our 2014 (1436 AH) it was 25 October, while in the year of writing, 2016 (1438 AH), it will fall on 3 October. In 2017 (1439 AH), it will shift back to 22 September. Only every 32.5 years will the dates of the festivals and New Year's Day in the Islamic year coincide. The Islamic and Gregorian calendars will not match, year for year, for another eighteen millennia, in the year AD / AH 20,874.

Each lunar cycle takes 29.5 days, so the lunar-Islamic year comprises 354 days with the months alternating from twenty-nine to thirty days throughout the year. This leaves every month about forty-four minutes short of the solar month, which rolls up to one full day every 2.73 years. Just as the Gregorian calendar has to add one extra day every four years, the Islamic world has to do likewise every three.

Each Islamic month begins a maximum of two days after the

new moon, when the crescent first manifests itself. Muhammad himself decreed that the first, seventh, eleventh and twelfth months were sacred, so it was unholy to wage war or engage in any form of fighting during their span.

The Islamic month best known to Westerners is the ninth, Ramadan, during which all but the frail, women who are either pregnant or breastfeeding, those already embarked on journeys, or those enduring diabetes must abstain from eating from dawn till dusk. The name means to burn, but it is unclear whether this relates to the prevailing weather (Ramadan usually falls in June or July) or for the fasting, perceived as burning bad habits and impurity from the bodies of the faithful.

THE HEBREW CALENDAR

Ill at ease with both the notion of starting their calendar at any man-made marker and also the BC and AD concept because they are still awaiting the Messiah, Jews embarked on the laborious task of totting up the ages of everyone mentioned in the Tanakh, or Hebrew Bible, to work back to a date of Creation. Designating their calendar scale Anno Mundi or AM, standing for the Year of the World (Creation), we are at the time of writing, in 2016, in the year 5776 which, just to confuse matters further, began at sunset on 13 September 2015 (Gregorian) and will terminate at sunset on 2 October 2016.

All Jewish days run from sunset to sunset, or the point at which the first three stars become visible in the heavens, if you want to get *really* picky. For practical purposes, modern Jews have opted instead for the point at which the sun is seven degrees below the horizon. Either way, all Jewish festivals begin the night before the day in question: their Sabbath, for example, starts at sunset on Friday evening. And the confusion does not stop there, as Judaism has no clock.

Each 'day' is reckoned to be the hours of daylight only, so the length of a Jewish day varies from winter to summer. Nevertheless, this span of daylight – whatever it may be – is still divided into twelve equal 'hours', the duration of which, like their days, will vary from month to month. Each of these hours is further divided into 1,080 halakim, or portions.

In the eighteenth century a considerable number of Jews migrated to northern Scandinavia where the 'day' can last a month in summer, as can the 'night' in winter. Needless to say, this played havoc with their attempts to judge the correct times of prayer and to properly position their festivals of significance.

EASTER AND PASSOVER

Both Easter and Passover are notoriously moveable feasts in that they are both tied to the occurrence of the vernal equinox, when the northern hemisphere starts to tilt towards the sun, signalling the start of spring. This can occur any time between

22 March and 25 April. For the Christian church this is, in the main, a pagan hangover from the old lunar calendar, with few in the UK aware of the Easter Act of 1928 as born of secular disenchantment with the inability of the Church to fix the festival to one specific Sunday. The act demanded that Easter Sunday be the second such day in April, which narrowed the window to the seven days from the ninth to the fifteenth day of that month. Although the act was passed, it still sits sulking on the statute books, as no administration since 1928 has been willing to risk the vote-losing fallout from starting a battle with the Church of England over its enforcement.

Passover presents its own problems. As is the date of Easter linked to the lunar cycle and the vernal equinox, Passover, which often coincides, commences on the full moon of the fourteenth day of the month of Nisan, the first month of the Jewish calendar, which has one foot in Gregorian March and the other in April. The Jewish name for Passover is Pesach, and 'paschal' can mean pertaining to either Easter or Passover, which within the borders of Israel spans seven days but for celebrants in other countries runs for eight days.

Allegedly, Passover was born of the Jewish slaves in Egypt daubing their doorways with the blood of paschal lambs, to make sure that the angels of death would 'pass over' them when they came to slaughter the firstborns in Egypt. This act of mindless savagery inflicted, the children of Israel got their marching orders in the year that Jewish tradition holds to be 2450 AM (1300 BC).

BC AND AD

This system of dating evolved from the labours of Dionysius Exiguus, or Dennis the Humble (c. AD 470–c. 544), an influential monk of Scythia in the Eastern Roman Empire, which approximately corresponds to modern Romania and Bulgaria.

In AD 525 Dennis was given the job of calculating the dates of all the paschal full moons and Easter Sundays for the period spanning AD 532 to 626. He had to work on a paschal calendar, which took what we now call AD 284 as its starting point, that being the year that the Roman Emperor Diocletian (AD 244–311) started his extremely brutal reign. As Christians had formed the main focus of Diocletian's homicidal spleen, Dennis was less than happy about using such a date as the basis of his calculation.

While Dennis' superiors promoted the date as the inauguration of the Era of the Martyrs, Dennis decided off his own bat that AD, which then stood for Anno Diocletiani, should be supplanted with Anno Domini, 'the year of our Lord'. Dennis avoided direct questions as to how he decided on the year of Jesus' birth, which he confidently proclaimed to be exactly 531 years before Year One on the scale of Anno Diocletiani. This he promoted as AD 1, as the mathematical concept of zero was unknown to him: Islamic and Hindu mathematicians had been using zero for centuries but the concept would not gain a foothold in Europe until the close of the sixteenth century.

Few took any notice of Dennis' suggestion, strangely preferring to remain faithful to the baseline of AD 284. However, AD and BC did start to gain traction in Christian circles with the newly converted Saxons imposing the system in Britain as early as the seventh century. Catholic Spain resisted until the fourteenth century, but some churches never took it to heart; the Egyptian Coptic Church, for example, still perceives AD to be Anno Diocletiani, which is why, for them, this year of writing (2016 minus 284) is 1732–3.

THE IMPERIAL YEAR ZERO

Until 1873, when it switched to the Gregorian calendar, Japan ran on the lunar-solar Chinese calendar it adopted back in the sixth century. But this changeover to Western measurement of time was not welcomed by all, especially not by the Japanese military, which continued to use the old imperial calendar dating from the mythical founding of their nation, the year the West designates 660 BC.

In 1940 the Japanese Air Force received its first delivery of a new fighter from Mitsubishi. In keeping with their habit of designating weapons by the last digit of the imperial year of their inception, as this was the year 2,600 on the imperial calendar they called it Zero.

After the close of the Second World War, the US occupation forces banned the imperial calendar, which now tends only to crop up in obscure legal documents with pretensions to antiquity.

NAME THE YEAR

It rarely occurs to observers of the Gregorian calendar that others march through the years to the beat of a different drummer. It may at the time of writing be 2016 for those who follow the Gregorian calendar, for many, but certainly not for all.

In the AUC calendric system 2016 is the year 2769. Standing for *ab urbe condita*, or the founding of the city (Rome), this system is used by scholars for dating events in the ancient world.

2016 is equivalent to 1465 in the Armenian Apostolic calendar, which began its count in AD 552, the year the Armenian Church split from Rome over theological disagreements.

According to Assyrians throughout the world, their official calendar renders 2016 as 6766, counting from 4750 BC when they built their first temple at Assur on the banks of the Tigris in modern Iraq.

For the eight million followers of the Baha'i faith, mainly scattered throughout India and the Middle East, 2016 is the year 173. Adherents start their calendar in the Gregorian year of 1844, when their first spiritual leader, or Bab, proclaimed he had been granted divine insight into the Koran.

Officially adopted by Bangladesh in 1987, the Bengali calendar makes 2016 the year 1423–4, as counted from AD 594 when the Bengali King Shasanka had a vision instructing him to follow Sanskrit texts on astronomy to construct a new calendar.

For the Berbers and Tuareg, still wrestling with their old Julian-based system, 2016 is the year 2966, as indeed it would be for many other countries had they not jumped ship to the Gregorian calendar.

On the British regnal calendar, as used for countless official documents in the UK, 2016 corresponds to 64–65 Elizabeth II, counting from the commencement of the reign of the present queen in 1952.

In the Buddhist calendar, 2016 is 2560, starting from the year the Buddha achieved parinirvana after his death in 483 BC. Most Buddhists assert that their leader attained such level of serenity in AD 544, but some factions argue for 543 or even 545.

For those in Myanmar, running on the old Burmese calendar, 2016 is the year 1378. With the New Year starting in April, this calendar has undergone repeated restructuring since its inception in 3102 BC, the last of which was imposed in AD 638 when the present count recommenced.

Despite Chairman Mao Zedong's (1893–1976) announcement on 1 October 1949 to adopt the Gregorian calendar, many traditionalist Chinese still use the calendar of the Yellow Emperor (the mythical emperor Xuanyuan Huangdi), whose reign reputedly began in 2698 BC, making 2016 the year 4714.

For many Japanese, 2016 is the year Heisei 28, counted from the death of the Emperor Hirohito (b. 1901) on 7 January 1989, he being the last emperor to be regarded as a living god.

TIME AND CLOCKS

The word clock strictly refers to a timepiece with a bell (from French, *cloche*). Before such devices the running of sand and water were used to mark time – but the daddy of them all is the simple sundial. Had these been first developed in the southern hemisphere, clockwise would denote rotation to the left. Sundials also explain why clock faces are traditionally round; true, they present problems around the equator and at the poles, but nothing is perfect!

However you choose to carve up the day, clocks have always been used to mark cultural trends rather than dictate them. For example, although noon is now twelve p.m., named from the Latin *nona* (ninth), this was for much of medieval Europe the hour we now call 3 p.m., as the working day started at 6 a.m., with the first break nine hours later. Plays and other entertainments put on in the local castle at 2 p.m. were called matinees, from the French *matin* (morning) but, as the working day became less arduous and noon moved back to its present position, matinees were marooned in the new afternoon.

In hotter countries the day still started at 6 a.m. but, due to the midday heat, the workforce rested six hours later, at 12 p.m. (noon). Termed the *sexta hora* (sixth hour) in Latin, this evolved into the modern 'siesta'.

DAYS, HOURS, MINUTES AND SECONDS

Yet another hand-me-down from the Babylonian obsession with astrology.

The Babylonians revered the numbers twelve, sixty and three hundred and sixty; the first two multiplied together gave the third, which was the number of days in their lunisolar year – the solar year comprised 365.24 days and the lunar year 354.38: add the two together and divide by two results in 359.8, this being rounded up to three hundred and sixty for convenience. In keeping with their notion of twelve signs of the zodiac, the Babylonian year was subdivided into twelve months.

Because they believed their lives were subject to the influence of their twelve zodiacal signs, the Babylonians decided that both the span of daylight and that of the night should be divided into twelve equal parts, which explains why we now have twenty-four hours to each day. Furthermore, to keep in step with their beloved number sixty, each hour was divided into sixty minutes and each of those minutes into sixty seconds.

The *ush* is the only Babylonian time unit we have not adopted. A unit of four minutes, it was most likely invented because each day comprised 1,440 minutes, which divided by four takes us back to three hundred and sixty. That said, the earth takes twenty-three hours and fifty-six minutes – four minutes short of twenty-four hours – to rotate on its axis and bring an observer back under the same star or constellation, but the Babylonians could not have worked that out – could they?

Because three hundred and sixty was thought to represent the circle of the years, in Babylonian maths the circle was likewise divided up into three hundred and sixty degrees and each of those degrees subdivided into sixty minutes.

THE SUN STANDS OVER THE YARDARM

On the old square-rig sails, the top horizontal timber (or yardarm) was used as a rough guide to judge the right time to pour out the first rum-issue of the day. Obviously, this would depend greatly on the ship's latitude and direction of passage but as a general rule in the North Atlantic, where the expression and practice originated, anyone standing on the main deck would see the sun standing above the top yardarm at about 11 a.m.

UP IN THE AIR

Another problem of human attempts to measure time arises from your position in relation to the planet – not in which time zone you stand, but your altitude.

In 2010 Dr James Chin-Wen Chou of NIST, the American National Institute of Standards and Technology, conducted experiments with an atomic clock. Accurate to one second in 3.7 billion years, this ran on the 'ticking' of a single ion of aluminium that 'ticked' between its two energy states one million billion times a second.

It has long been known that time accelerates the deeper into space you go, as you are moving further and further away from any gravitational influence. Gravity generates a form of time dilation, so the further away you move from the point of such force, its increasingly weakened influence allows time to pick up speed. So, a person on top of a mountain not only weighs less than they do at the foot of the peak, they also age fractionally faster. However, Dr Chou was able to measure a minute acceleration of time when the clock was raised a mere twelve inches off the laboratory floor. This variance was checked and double-checked to ninety billionths of a second over seventy-nine years so it is unlikely to have a profound effect on anyone's lifespan, but it can now be said that your head ages faster than your feet and that going upstairs to bed will shorten your life.

It had also been known for decades that time slows as you approach the speed of light. This is known as the Twin Paradox. If one twin gets into a rocket and accelerates away from earth, at or beyond the speed of light, for a while before returning in the same fashion, he or she will be younger than the sibling on the ground.

But Dr Chou proved that you do not need rockets or warp drive to experience the effect. When transported at a mere twenty miles per hour (32 km/h) the clock slowed by several billionths of a second. So, drive slower and live longer.

SATNAVS AND TIME DILATION

For those of us with our feet on the ground such variance in time caused by altitude and speed are mildly interesting but of no discernible impact, but that cannot be said of GPS satellites feeding data to the irksomely monotone satnavs in our cars.

As mentioned above, the faster you travel in space the slower time passes and the further you are from the earth's gravitational pull the faster the passage of time, so positional clocks on a GPS satellite are subject to these two opposing influences. Due to a satellite's speed alone – 8,700 miles per hour or 14,001 kilometres per hour to keep pace with the earth's rotation – time runs slower by 7,200 nanoseconds every day. Its altitude of 12,000 miles means that there is the counter-indication of time passing 45,900 nanoseconds a day *faster*. Deducting one from the other leaves the satellite's balance 38,700 nanoseconds faster every day.

Although such miniscule variance is unlikely responsible for our satnavs carefully leading us into dead ends and farmyards, it does have to be accounted for, as the roll-up over several years would have a discernible effect.

GEOSTATIONARY ORBIT

In the general perception GPS satellites stand in a fixed and geosynchronous position above a single spot on the earth's surface whereas this is only possible directly above the equator. In reality there are twenty-four GPS satellites, each of which 'rise' and 'set' twice a day. Divided into groups of four, each group travels one of six separate orbits so that eight are visible from any position on earth at any one time.

Far from our dashboard satnavs being guided – or misguided, as the case may be – by a single satellite directly above, information is being constantly fed to a GPS receiver by the visible eight. This data is combined and averaged out to feed satnavs a reading accurate to about ten metres.

DECIMALIZATION OF TIME

The first to challenge the time structure imposed on us by the six-mad Babylonians were the revolutionary French, who decided to decimalize both the clock and the calendar, launching such lunacy in 1793, which was rebranded year one.

Try as they might, the members of the committee appointed by the Revolutionary Council to oversee the change could not equally divide three hundred and sixty-five days to come up with a ten-month structure, so they opted instead for twelve months of thirty days, each divided into three ten-day blocks called decades. The first day of each decade was a day of rest, which did not go down well with the working populace, who hankered after the old Gregorian structure in which they got one day off in seven.

The months were renamed according to their prevailing weather, such as Brumaire (mist) and Thermidor (heat), and the days of the decades numbered consecutively, moving towards a more scientific and secular system. The first day of this new-style year, in 'Grape-harvest' month, fell on 22 September 1792, the date of the founding of the First French Republic.

As for the clock, this was redesigned to have a face of ten hours, each of one hundred minutes comprising one hundred seconds. It is fair to say that the only people pleased with the new system were calendar printers and clockmakers, who made a fortune. These same people would again make a killing in 1805, when the new Emperor, Napoleon I, quickly reverted to the Gregorian calendar.

TIMEKEEPING

Responsible for setting the precedent of circular clock faces, the sundial could be extremely accurate, with the Egyptians erecting them on a massive scale in the form of obelisks, as early as 3,500 BC. These first ever public clocks cast their shadows across hour markers built into the paved squares in which they stood, with Cleopatra's Needle on London's Victoria Embankment, its twin in New York's Central Park, and the Luxor Needle in Place de la Concorde, Paris, being three surviving examples of these gigantic timepieces.

As sundials were pretty useless after dark, the ancients tried to set up moon dials. These proved to be more trouble than they were worth, as they only worked on the night of the full moon. For every night of the fortnight following the full moon, such 'clocks' would read about fifty minutes slow, and for each night of the two weeks leading up to the next full moon the dial would be about the same number of minutes fast. So, in the week either side of the full moon, a moon dial is about five-and-a-half hours slow or fast. Useless perhaps, but these dials provided the first proof that the orbit of the moon was anything but circular.

For a sundial to give accurate readings it must be properly aligned, which means that in the northern hemisphere the dial's shadow-caster, or gnomon, must point to true south, and to true north in the southern hemisphere (where the dial will read anti-clockwise). On or near the equator, they are a nightmare to set up, with the gnomon requiring to be aligned with the direction of the earth's rotation.

WATER CLOCKS

As stated above, sun- and time-obsessed Egyptians were frustrated by their inability to tell the time at night and, having tried the abortive moon dial, invented the water clock.

Around the middle of the sixteenth century BC we find the first reliable mention of a water clock, in the ancient temple complex at Karnak in Luxor, Egypt. There are claims of functioning water clocks existing as early as 4,000 BC in India and China, but these cannot be substantiated.

Calibrated by a sundial, the first examples were, like all good ideas, blissfully simple, with the flow from a main reservoir being regulated to fill a receiver with graduated markings up the inner surface. Still having to be calibrated by sundial, the water clock rapidly gained popularity over its predecessor as it became more complex.

The Babylonians were quick to realize that with scales to weigh the increasing weight in the receiver and gears to turn a dial you could have something very close to the modern clock. The Greeks and Romans refined things further by extending the use of gearing and including the first escape mechanism, so by the fourth century BC water clocks featured alarms, little doors opening on the hour to reveal various gods – the lot. It was this second refinement that would be copied by German clockmakers of the 1620s to produce the first cuckoo clocks, later hijacked by the Swiss.

Both Greek and Roman courts used water clocks to restrict

the time for which anyone could speak, and all Athenian brothels used water clocks to call time on clients. In fact, water clocks were so good that it is fair to say that come their replacement across seventeenth-century Europe by the admittedly more accurate pendulum clock, the only difference was the method of drive, from water weight to a metal spring: the complicated gearing remained much the same.

SANDGLASS

 Hourglass is a slightly misleading term in that the contents could be calibrated to any span of time. Nor for that matter were they routinely filled with sand, a material far to coarse and angular for smooth transfer from one bulb to the other, and too inconsistent of grain size to give an accurate reading. In Renaissance times the better examples were typically filled with marble dust harvested from the masons' workshops or pulverized eggshell, and although sand is today sometimes used for decorative hourglasses, this tends not to be beach sand but the kind of finer-grain sand found on riverbeds.

First noted in Alexandria in 150 BC, the hourglass did not find its way into Europe until the fourteenth century when it was first popular with navigators irked by condensation and motion-induced inaccuracy of their water clocks at sea. No surprise there, so the hourglass was the first means by which navigators could calculate longitude with reasonable precision.

The device found equal favour on land, for everything from measuring cooking times to clergy being obliged to turn one in the pulpit at the beginning of a sermon to prevent them droning on ad infinitum. In fact, they are still put to similar use in the Australian Parliament in Canberra, which uses either a four-minute glass or a two-minute glass to goad members into making their minds up quickly as to which way they intend to vote by joining the appropriate lobby in the event of a division.

INCENSE CLOCK

More a feature of Eastern timekeeping, these comprised a lantern housing an incense stick of calibrated dimensions with wire-and-bell markers at the hour or divisions thereof. As the stick burned down, the thin wire would break to allow its bell to drop into a copper drum below, giving an audible alert of the passing time to those present.

Until 1924, when cost-conscious clients demanded that the duration of their visit was timed by modern clocks, this remained the standard Japanese method of working out the fee owed a geisha for his or her attendance, depending on the number of bells on that drum-plate or complete incense sticks consumed if in attendance for longer periods.

Entertainers of great artistic skill and education, geisha are much misunderstood in the West, where there is no equivalent. Female geisha were unheard of until the middle of the eighteenth century when the first few provoked mixed reactions in Japanese culture: not until the turn of the nineteenth century did female geisha start to outnumber their male predecessors. Male geisha still function in Japan, but are more usually called *taikomochi*.

TIME IN YOUR POCKET

The increasing demand for timekeeping on the move forced clockmakers towards portable timepieces for the night watch, a kind of early police force in cities, and for those on watch duty at sea – hence 'watch'.

Timepieces intended to be worn first emerged in sixteenth-century Germany, but these were so cumbersome that they had to be hung about the neck by a rather strong chain. We do know that Robert Dudley, first Earl of Leicester (*c.* 1532–88), presented Elizabeth I of England with a watch 'to be worn about the arm' in 1571, and the golden rule seems to have been wristwatches for women and pocket watches for men, for whom the waistcoat was developed in the 1670s as a vestment with small pockets in which to keep their timepieces.

The wrist/pocket gender divide arose from the fact that small timepieces were until quite recently prone to fouling in damp weather and, generally, men followed outdoor pursuits while ladies stayed at home. Wristwatches remained decidedly unmanly until the 1880s, when British Army officers became the first men to wear them to enable them to synchronize attacks on horseback in conflicts such as the Boer War (1899–1902).

These military timepieces were basically pocket watches mounted on a strap, but the creeping barrage tactics of the First World War (1914–18) created the demand for purpose-built, robust wristwatches. Artillery and infantry officers very much needed to know the precise timing of each other's actions to reduce appalling levels of casualties due to so-called friendly fire.

THE CLOCK WITH NO HANDS

Although the Romanian chronobiologist Franz Halberg (1919–2013) coined 'circadian' – rhythmic, biological cycles recurring at approximately daily intervals – in 1958, people had for centuries noticed that all living things went through a repeating pattern of behaviour in the course of twenty-four hours.

With its main triggers being dawn and dusk, the circadian clock in humans seems to re-set at 4 a.m. when body temperature is at its lowest, perhaps giving rise to the myth that more people die at such time than in any other hour in the day (11 a.m. is the real Hour of the Wolf). Either way, the importance of the circadian clock being able to re-set itself according to the rhythm of light and dark is responsible for the jet lag experienced by people travelling so fast and so far that they land before they have even taken off, according to the local time of their destination.

An altogether more severe demonstration of circadian disruption is being studied in Greenland, which has the highest known suicide rate in the world. You would have thought that the three-month nights would be the major cause of depression, but it is the relentless periods of daylight that coincide with peaks in suicides and homicides.

To prove the predictable reliability of circadian rhythm, eighteenth-century Swedish botanist Carl Linnaeus (1707–78) laid out the design for a flower clock, using clusters of plants known to open and close at precise times of the day. His design incorporated around fifty different flowers, ranging from goat's beard (*Tragopogon pratensis*) that opens around 3 a.m., to the daylily (*Hemerocallis*) that closes around 7 p.m. Had Linnaeus actually planted this device, it would have proved largely accurate.

TIME SIGNALS

Prior to the advent of radio and its familiar pips on the hour, cannons were the first time signals heard in every city and port. Typically fired at noon, these allowed ships' captains to set all timepieces used to calculate longitude prior to slipping anchor. That said, Vancouver installed a gun in 1894 to be fired at 6 p.m. on Sundays to call for a cessation of all fishing: it still fires nightly at 9 p.m., but only as a time signal.

Other Canadian guns, such as those installed in Halifax and Quebec City, still fire at noon, as do those in Cape Town and Santiago in Chile. The Roman gun, introduced in 1847 by Pope Pius IX (1792–1878) to coordinate the ringing of church bells in the city, still fires at noon on the Janiculum, a hill on the west bank of the Tiber.

With light, as a general rule, travelling faster than sound, cannon were replaced by visible signals such as the time ball, invented in 1829 by the British Admiral Robert Wauchope (1788–1862). A brightly coloured metal sphere released at noon to drop down the pole on which it was mounted, the time ball on top of the Greenwich Observatory still drops at noon (Greenwich Mean Time) every day, a device copied by New York since 31 December 1907, the date of the first Times Square ball drop to signal the New Year.

LOCAL TIME

With the passing of time being an abstract concept which everyone tried to measure by the position of the sun at noon, the imposition of a standard system across an entire country or even continent made the task of getting an angry cat into a shoebox look easy.

Russia, for example, has eleven time zones ranging from GMT+2 to GMT+12, with a time difference spanning ten hours. China had five time zones that were abolished in 1949 by the fledgling People's Republic, which decreed the whole country would run on Beijing Time, equating to GMT+8. Even on an island as small as the UK, until 1880 there was no straight answer to 'What time is it?'

As in every other country in Europe, every town in early nineteenth-century Britain was running on its own local time, dictated by the zenith of the sun in its own back yard. Few journeyed very far in such times and coach trips were so slow that carriage clocks could easily be re-set, or ignored completely, so there were few time-travelling problems to contend with – but then came the railways.

Parts of the UK were anything up to half an hour out of sync with GMT, so when the railways started to move people about the UK at speed, problems did arise which were not much helped by the various railway companies agreeing between themselves to run on GMT, leaving their timetables hopelessly out of sync with local times of arrival and departure.

Such matters came to a head on 25 November 1858 at the Dorchester Assizes, which was convened to hear the case of *Curtis v. March* at 10 a.m. that day – but the court clock was set to GMT and the town clock to local time. When the defendant and his lawyer failed to appear at 10 a.m. GMT, the bench found in favour of the plaintiff. March appealed to have that decision overturned by a higher court, which ruled that 10 a.m. meant local time and not this irksome railway time, as GMT was then popularly known due to having been first employed by Great Western Railway.

This ruling remained in place until 1880 when GMT was officially adopted as a blanket timescale for the UK. All except the British Royal Family, that is, who ran on what was called Sandringham Time – thirty minutes ahead of GMT.

SANDRINGHAM TIME

With male members of the British Royal Family passionate about hunting on the Sandringham estate, all the clocks were set half an hour fast to conserve daylight for field sports. But after this confusion of clocks led to the 'murder' of King George V (b.1865) in 1936, the royal observance of Sandringham time was abandoned.

George was certainly on his last legs at Sandringham on the night of 20 January that year and his attending physician, Lord Bertrand Dawson (1864–1945), was anxious that he die in time to make announcement possible in the morning edition of *The Times* instead of 'less appropriate evening journals'.

Knowing that newspaper's deadlines for last-minute alterations to the front page and confused by his being surrounded by clocks giving the wrong time, Dawson thought it time to hurry things along by giving the king a lethal overdose of morphine so he could die in time to make the right headlines. Before he acted, Dawson phoned his wife in London and told her to ring *The Times* and tell them to hold the front page.

Later that same year, Dawson opposed a motion to legalize euthanasia brought before the House of Lords which, in the light of the above revelations of his personal diaries made public in 1986, smacks of hypocrisy to say the least.

GMT

Although many presume GMT to be a super-accurate measure of time, it is no such thing, as the M stands for 'mean', as in average. GMT noon is only 'right' once a year. Because the earth's orbit is elliptical and of variable speed as the planet also shifts on its axis, GMT noon at the Greenwich Observatory can be anything up to sixteen minutes before or after the point at which the sun stands in its zenith above the Greenwich Meridian at noon.

As the GMT day marks the time it takes for the earth to rotate from noon to noon it found little favour in the scientific

The Royal Observatory, Greenwich

community which, in 1935, opted for Universal Time which marked out its days from midnight to midnight. In 1972, Universal Time was further refined into Coordinated Universal Time, which included the insertion of leap seconds to allow for the fact that the earth not only wobbles on its axis but is also slowing down its rotational speed due to gravitational drag. The most recent, thirty-sixth, leap second was added at midnight on 30 June 2015.

All atoms vibrate and, in 1967, it was established that an atom of caesium, at its atomic level, did so at the rate of

9,192,631,770 times per second to generate the new standard for keeping Coordinated Universal Time with atomic clocks. Today, the quoted International Atomic Time is the arithmetic mean of readings taken from four hundred atomic clocks spread across fifty countries. As the eccentricities of the earth's orbit cause the year to gain or lose the odd nanosecond here or there, in the rarefied atmospheres of astronomy and astrophysics a year is no longer 365.24 days but 290,091,200,500,000,000 caesium oscillations long – but who's counting?

GREAT PARTY, WRONG YEAR

In the April of 1997 an atomic clock was set up by Accurist on the meridian at Greenwich to give the public a visual display of the thousand-day countdown to the turn of the second millennium.

This was not the only such gimmick but the various sponsors could not agree over the inclusion of leap seconds. Some had them built into the programming while others decided to refrain from their inclusion until they were actually inserted by Coordinated Universal Time. So, at any one point, the millennium Countdown clocks differed by anything up to three seconds.

Be that as it may, when the Accurist display at Greenwich flicked over to a bank of zeros at the turn of 31 December 1999 to 1 January 2000, those in attendance popped their champagne corks to toast the new era, blissfully unaware that the clock they had been so avidly watching was slightly out (by exactly one year) due to the mistake made by Dennis the Humble – remember him?

As explained previously (see page 110), Dennis was the monk who created the AD and BC system in the sixth century after picking a year at random and pronouncing it to be the year of Jesus's birth. But, back in the sixth century no one in Europe had any knowledge of the mathematical concept of zero, so he called it year one. From one to one hundred is only ninety-nine, and so forth, so the new millennium did not begin at the turn of 31 December 1999 to 1 January 2000; it began on the turn of 31 December 2000 to 1 January 2001. But what the hell – it was a great party.

TEMPERATURE

Before looking at the development of the various thermometric scales it is perhaps best to dispel the pervasive myth that heat rises. The property of heat is a form of energy that will distribute itself throughout its environment in an even pattern, flowing from areas of high pressure to those of low pressure. If you aim a blowtorch at the centre of a metal plate for long enough, you will end up with a fair approximation of the Japanese flag: the heat will not only rise.

Neither, it can be argued, does heated air rise. Air expands with heat to become apparently lighter but, as the cold air surrounding a heated parcel of air in, say, a hot-air balloon, is heavier, it is subject to a greater gravitational pull downwards, forcing heated air upwards. Thankfully, for all who live on this planet, the reverse is true for water. Lower down the temperature scale, warm water is indeed lighter than cold, which is why radiators, on activation of the heating system, heat from the top down. But, at and below four degrees Celsius or 39.2 °F, cold water starts to expand to become lighter and, in its ultimate cooled form of ice, floats on warmer water. If ice sank we would all be in a lot of trouble.

FAHRENHEIT

The oldest of the mainstream thermometric scales, this was devised by the Polish-born Dutch physicist Daniel Fahrenheit (1686–1736) in Danzig during the unusually severe winter of 1708–9, resulting in his scale being arbitrary in the extreme.

Fahrenheit's first temperature marker was to set the normal freezing point of water, for which he nominated thirty-two degrees – some saying that this was inspired by his involvement with Freemasonry, in which there are thirty-two levels of enlightenment. Working back from this, Fahrenheit established his point zero at the temperature of the ice in Danzig that winter which, mixed with brine, just happened to equate to −17.8 °C on the yet to be invented centigrade / Celsius scale. His next marker was the temperature of his wife's armpit, which he used to establish one hundred degrees.

After his death, the Fahrenheit scale was recalibrated to mark the upper point at 212 °F, the boiling point of water at sea level, and to then move back in more precise steps to its freezing point of thirty-two degrees. The minor adjustment to each individual degree-step moved the temperature of Frau

Fahrenheit's armpit down from one hundred degrees to the now much-quoted 98.6 °F, the point of perfect health.

To convert Fahrenheit to Celsius you have to subtract thirty-two, multiply by five and then divide by nine. It was, and still is, a clumsy scale, which is largely ignored by the scientific and medical worlds. The only people still using it are American weather presenters.

CELSIUS

Formerly known as centigrade, this temperature scale was devised in 1742 by the Swedish polymath Anders Celsius (1701–44), who seems to have got things a bit base over apex.

Celsius was studying the effect of cold, which he mistakenly thought to be a similar but opposing force to that of heat; he never realized that what we call cold is simply an explanation of the absence of heat. Because of this, his original scale had zero as the boiling point of water and one hundred degrees as its freezing point.

Had he lived, Celsius would doubtless have revised his thinking, but tuberculosis carried him off two years later, leaving it to his best friend, Carl Linnaeus, the Swedish botanist regarded as the father of taxonomy and the founder of modern ecology, to put matters right.

Although Linnaeus did indeed turn Celsius' scale the right way up, he did not do this because he thought his friend had

ANYONE FOR CRICKET?

Many people know how to use an analogue wristwatch as a compass (for those who don't, with twelve o'clock to the left and the hour hand aligned with the sun, south is indicated halfway between the two) but you can also use one to measure the environmental temperature – providing you can hear crickets.

In 1897 the US physicist Amos Dolbear (1837–1910) published a paper on the correlation between ambient temperature and the chirping of crickets, a noise that many wrongly believe to be caused by the insect rubbing its back legs together. In an action called stridulation, the cricket actually grinds one serrated wing against the other, and does so in increasing frequency the higher the ambient temperature.

According to Dolbear's Law, as it is known, all you have to do is to count the number of cricket chirps in fourteen seconds and then add forty to that figure to arrive at the temperature in Fahrenheit. If you want the figure in Celsius, count the chirps in twenty-five seconds, divide that number by three and add four. It is remarkably accurate.

been wrong. Linnaeus was preoccupied with plants and their conservation, so he knew that freezing point was the degree at which they all started to suffer.

KELVIN

Having long expressed the need in scientific circles for an absolute thermometric scale, in 1848 William Thomson, later Lord Kelvin (1824–1907), announced his new scale that took absolute zero as its starting point.

Yet to be achieved in laboratory conditions, although some have come very close, the point of absolute zero is said to be that point at which all molecular activity stops – or appears to stop – which would indicate that the subject matter has no more heat to release. Although some argue that absolute zero is the point at which molecular activity is at its minimum and that it is impossible for all molecular activity to grind to an absolute halt, zero in the Kelvin scale equates to −273.15 °C or −459.67 °F. Either way, you'd need weapons-grade thermal undies to go out in that.

For his next defining point, Lord Kelvin took the triple-point of water, or the point at which the elements of gas, liquid and solid are present in equilibrium (which, on his scale, is achieved at 273.16 kelvins or 0.01 °C / 32.018 °F). This is the only scale not to make mention of degrees, with each kelvin defined as being 1/273.16th of the triple point of water.

THE TEMPERATURE OF CREATION

Although Kelvin had a mind to be reckoned with, when he tried to use temperature as a guide to the age of the earth his thermometric research eventually led him into a blind alley.

Falsely reasoning that earth had cooled at a consistent rate from its inception as a red-hot ball, and unaware that radioactive decay generates additional heat, Kelvin confidently announced that to have cooled to its current temperature gradient earth would have to be no less than twenty million years old and no more than forty million. Even had he known about the heat generated by radioactive decay, Kelvin's calculations would still have been wildly off the mark because he falsely presumed heat to be transported with uniform efficiency throughout the entire depth of the planet.

Even after it was established that this was far from the case, Kelvin refused to accept facts and stuck resolutely to his guns, all of which were pointing the wrong way. As explained in Mario Livio's book *Brilliant Blunders* (2013), Kelvin was so used to being right that he was not prepared to recognize the possibility of his being wrong.

Even after 1905, when the US physicist Bertram Borden Boltwood (1870–1927) proved beyond doubt that there were surface rocks older than five hundred and seventy million years, old Kelvin kept his hands over his ears while humming loudly.

If it matters, the earth is about 4.6 billion years old.

LIGHTNING

It is popularly stated that the core temperature of a lightning flash is six times that of the surface of the sun but, obviously, no one has been daft enough, let alone quick enough, to test this with a thermometer.

As lightning passes through the atmosphere it super-heats the gases to such an extent that it breaks apart the molecules and strips all the electrons from the atoms. Using an instrument called a spectrometer, scientists can identify the various bands of colour thus generated, some of which extend into the ultraviolet range, to arrive at a fairly accurate idea of the attendant temperatures, which can reach 30,000 k / 29,726 °C / 53,540 °F.

Although it is fair to say that the temperature at the core of a lightning flash is six times that of the surface of the sun, this is largely an exercise in semantics. Counter-intuitively, the surface of the sun is its coolest zone, at about 6,000 k / 5,726 °C / 10,340 °F. The temperature a few inches above the surface leaps to over 500,000 k / 499,726 °C / 899,539 °F. At its core, the sun is estimated to be over 15,000,000 k / 14,999,982 °C / 27,000,000 °F.

SEVEN STRIKES AND YOU'RE OUT

Those who say lightning never strikes in the same place twice have obviously never heard of Roy Sullivan (1912–83), one-time ranger of Shenandoah National Park, Virginia, USA. Depending on where you live and what you do for a living, the odds of being struck by lightning range from about 300,000:1 to 1,000,000:1, so the odds against Sullivan being hit seven times must be astronomical.

Between 1942 and 1977, Roy Sullivan required medical treatment for no fewer than seven lightning strikes to his head and upper body that variously burned off his hair, blew out his fillings or set fire to his footwear. The strike of 1977 hit him on the head while he was fishing on a clear day. He woke up just in time to fight off a bear attracted by the trout he had caught.

ORIGINS OF THE THERMOMETER

Scales of temperature are useless in the absence of calibrated devices by which to judge them, and the thermometer was a surprisingly late development.

The forerunner of the thermometer was the thermoscope, an open-topped but uncalibrated glass tube partially filled with a fluid known to expand under heat. Developed in medieval Italy, where they were filled with red wine, it is frequently said that Galileo was the first to produce and use a thermoscope, but it is far more likely that this was achieved by his contemporary, Santorio Santorio of Padua (1561–1636).

As a thermoscope could only give comparative readings – for example, this fluid is hotter than that one – Santorio packed his device in snow to mark the lower level, then held it over a candle to mark the maximum expansion. He then divided that span into 110 segments and put the thermoscope

into the mouths of subjects with fever to prove the connection between disease and body temperature. Santorio's oral thermoscope was the first clinical thermometer – not bad for 1612.

The first liquid-filled and enclosed tube was made sometime between 1652–54 by Ferdinand II, Grand Duke of Tuscany (1610–70), who used dyed alcohol as the reader fluid. He calibrated his device with three hundred and sixty incremental steps, as that was the number of degrees in a circle. Thus we have Ferdinand to thank for the word 'degree' transferring from geometry to thermometrics.

At the beginning of the eighteenth century, Isaac Newton found he got more accurate readings using linseed oil as the reader, and although others before him had used mercury, Daniel Fahrenheit was the first to produce a mercury-filled thermometer with a standardized scale in 1714.

The steady and reliable expansion of mercury

made it the ideal reader but more recent concerns over its toxicity and the health implication of one snapping in a patient's mouth have forced its substitution by other mediums.

HEAVEN VERSUS HELL

Mark Twain was fond of quipping he'd choose heaven for the climate but hell for the company, only to be proved wrong in 1920 by Dr Paul Darwin Foote (1888–1971), US pioneering expert in pyrometrics. Heaven, it seems, is by far the hotter of those alternative and final destinations.

Taking Revelation 21:8 at its word, that hell abounds with lakes of molten brimstone, now better known as sulphur, Foote established the ambient temperature of Old Nick's domain to be below 444.6 °C / 832.28 °F, as that is the point at which liquid sulphur turns to gas. Isaiah 30:26 informed Foote that in heaven the light of the moon would be as that of the sun, while the sun itself would also radiate seven times the light of seven days.

Applying the Stefan–Boltzmann fourth power law of radiation, Foote came up with $(H/E)^4 = 50$, which fixes the temperature of heaven at 525 °C / 977 °F, which is 81 °C / 177.8 °F hotter than hell.

THE MPEMBA EFFECT

It is frequently asserted that hot or even boiling water will freeze faster than an equal volume of cold. While this counter-intuitive concept has been discussed by notables ranging from Aristotle to Descartes, it only holds true for very small quantities and in quite specific circumstances: it is impossible for a bucket of boiling water to freeze faster than a bucket of cold.

Water that has been heated or boiled and allowed to cool will always freeze faster than an equal, cold sample as the act of boiling removes impurities and tiny air bubbles that would otherwise impede freezing. It is also possible for a small sample of hot water to freeze faster than a cold sample if placed in tall and narrow containers. Descartes postulated that this is due to the hot sample 'stirring' itself, as chilled water increases in weight and repeatedly falls to the bottom throughout the freezing process, whereas the cold sample has to wait for its entire mass to be uniformly reduced in temperature. *Agito ergo sum* ('I move therefore I am'), the great man doubtless exclaimed.

The phenomenon got its name in 1963, when Tanzanian schoolboy Erasto Mpemba was making ice cream at the Magamba Secondary School and shoved his still-hot mix into the freezer, only to find it the first to set. Although several universities are conducting research into the mercurial phenomenon, none has yet been able to identify the precise circumstances under which the Mpemba Effect can be routinely and reliably demonstrated.

HOT OR WHAT!

Most people have heard of absolute zero (the lowest theoretical temperature, at which atoms would stop moving: 0 k / −273.15 °C / −459.67 °F), but few outside the world of physics and particle physics have heard of the conjectured point of absolute hot, at which matter holds its maximum amount of heat and molecular activity.

Early in 2012 in New York, Brookhaven National Laboratory's 2.4-mile-long Relativistic Heavy Ion Collider smashed particles of gold into each other to produce, if only for a split second, a temperature of 4 trillion °C or 7.2 trillion °F, which is 250,000 times the temperature at the core of the sun. Later that same year, the better-known Large Hadron Collider conducted a similar run with particles of lead to achieve a split-second reading of 5.5 trillion °C or 10 trillion °F.

To be fair, the LHC does have an advantage over Brookhaven in that it enjoys a circular run of seventeen miles or twenty-seven kilometres, but both fall way short of the conjectured point of absolute hot, which involves thermal equilibrium at 7.56e+32 Fahrenheit – or 420,000,000,000,000,000,000, 000,000,000,000 kelvins, and if you want to convert *that* to Celsius, just stick a 1 at the front.

THE CALORIE

The calorie is a much-misunderstood unit which was first defined in 1824 as a unit of heat by the French physicist Nicolas Clément (1779–1841) as the amount of heat/energy required to raise one gram of water through one degree Celsius at one atmosphere. Everything containing energy contains calories, not that you would want to eat it. One ton of coal, for example, holds 7,004,684,512 calories awaiting release through combustion.

We all need to derive energy from food but the weight-conscious should be aware that food labelling too is misleading in that the food-manufacturing lobby, aware of its current reputation, chooses to fudge the issue by presenting kilocalories as calories. So, when the nutritional information on a bar of chocolate states that it contains but three hundred calories, this should be understood as three hundred kilocalories, or three hundred thousand calories.

IF YOU CAN'T STAND THE HEAT...

In 1912 US chemist Wilbur Scoville (1865–1942) was working on a chilli-based cream to treat arthritis and decided that he first needed to grade the heat of the various chillies he was using. He devised a scale based on the equal amounts of sugar water required to dilute the sample extract to such degree that the taster felt no pain.

In incremental steps of one hundred Scoville heat units, Wilbur Scoville rated all known chillies of the day, from the sweet bell pepper at zero up to pure capsicum at sixteen million SHU. Although the principle of his scale is a bit shaky as it depends on the individual tolerance of his tasters, it is a good guide and still in use. The pepper sprays used in America as a non-lethal response by police, and by hunters wishing to deter inquisitive bears, deliver about 5,000 SHU.

Original red Tabasco sauce rates between 4,000 and 5,000 SHU depending on the batch, but the hottest product available was Blair's 16 million Reserve, a name that says it all. Not a sauce but pure crystals of capsicum, these are about 8,000 times the strength of Tabasco. At the time of writing, the hottest chilli per se is the appositely named Carolina Reaper, with 1,569,300 SHU.

And it seems old Wilbur was right about chillies' medicinal value. Capsicum has now found a role in topical analgesics and, in tablet form, is also used to treat intestinal problems, high cholesterol, nerve pain, malaria, and even to instigate weight loss in the obese.

SOUND AND LIGHT

Somewhat depressingly, man tries to weaponize anything at his disposal, with both light and sound having been pressed into such service. Laser-guided missiles and lethally accurate red-dot gunsights are common, while sonic cannons are now a reality. Used for crowd control and in riot situations, low-frequency sound waves can induce nausea and higher pulses can even disrupt vision and burst eardrums.

As people over the age of twenty lose the ability to detect sound in the higher frequencies, high-pitched devices are now employed in shopping centres to deter teenagers from hanging around in menacing gangs. 'Screamers' are incorporated into modern burglar-alarm systems to make it impossible for any intruder to remain on the premises – so not all bad, then.

Sound differs from light in that it has no universal constant when it comes to its measurement in terms of speed. Light enjoys a constant of 1,080 million kph or 671 million mph if travelling in the vacuum of space (where sound cannot travel at all). So it is true: no one can hear you scream in space. And at the beginning of time, the Big Bang – if it happened at all – was silent.

THE CRACK OF CREATION

'The Big Bang' was coined to ridicule the theory that the universe began with one cataclysmic and explosive expansion from nothing.

The British astronomer Sir Fred Hoyle (1915–2001) was a lifelong adherent of the theory of a steady-state universe with no beginning and no end. So when the Vatican-backed priest and astronomer Monsignor Georges Lemaître (1894–1966) proposed the creation of the universe by one cataclysmic event, Hoyle felt that this was an attempt to impart scientific spin to the notion of a Divine Creation in order to validate the Vatican standpoint.

Speaking on BBC Radio 3's *Third Programme* on 28 March 1949, Hoyle spoke of the Big Bang theory in ridicule, only to see the expression adopted by Lemaître and his followers. Perhaps Hoyle was right: at the time of writing, the theory is under increasingly vociferous attack by numerous astrophysicists of international reputation.

MAN MEASURES SOUND

It was the advent of early muskets and cannon that alerted people to the fact that there was something odd about sound. When observed from a distance the puff of smoke or muzzle-flash on firing could be clearly seen long before the sound was heard. While it is true that, long before the advent of firearms, all and sundry were aware of the delay between seeing lightning and hearing thunder, as none then knew that the former causes the latter, no connection was made.

The first to try to measure the speed of sound was the French polymath Pierre Gassendi (1592–1655). In 1635 he positioned cannon at one thousand metres' (3,281 ft) distance with a crew instructed to fire their guns on hearing the pistol fired by an assistant at Gassendi's observation point. Yes, it is easy to spot the flaw in the great man's plan. Gassendi also

failed to factor in the reaction time of both himself and the gun crew, so his resultant figure of the speed of sound being 487 m/s or 1,070 mph was way too high.

In 1864 the French scientist Henri Regnault (1810–78) arrived at a far more accurate figure by eliminating the human element from the equation. His apparatus was a rotating cylinder of soot-covered paper against which a metal stylus was held in place by a light spring. Attached to the stylus were two wires; one wire trailed one thousand metres to be taped across the muzzle of the cannon, while the other ran through a sound-sensitive diaphragm on the test bench next to the cylinder.

When the gun was fired the first wire broke to make the stylus jump, and when the sound broke the diaphragm the stylus jumped again. Knowing the distance to the cannon and the speed of the cylinder's rotation, Regnault came up with a speed of 335 m/s or 750 mph, which is pretty close to the now accepted figure of 343.2 m/s or 768 mph for the speed of sound at sea level and on a dry day.

MYTH OF THE GUN-SILENCER

We have all seen the cinematic assassin screw a cigar-sized cylinder to the muzzle of his gun to reduce the sound of death to a muted 'phut' – but such devices only work in Hollywood.

Modern ammunition is supersonic, so no device fitted to the muzzle is going to affect the sonic boom it creates after leaving the gun. The average gunshot generates, albeit for a fraction of a second, about one hundred and sixty decibels – the same as standing next to a 747 jet engine – and a 'silencer', more properly called a suppressor, will only drop that to about one hundred and forty decibels. To put that into perspective, it is about the same as putting your ear to a police siren.

Suppressors are used by snipers to eliminate muzzle-flash and to transform the noise of their gunfire from a sharp retort to a flattened sound which is more difficult to pinpoint. They are also used by SWAT teams to make shooting possible in confined spaces without rupturing everyone's eardrums and rendering them deaf for life.

HOW FAST IS SOUND?

Sound needs a medium through which it can travel, and the denser the medium the faster it goes. At sea level and on a mild day, sound will happily trot along at 768 mph / 1,236 kph but if the temperature increases it will speed up considerably as the molecules become more fluid and thus better able to transmit the sound. So on a warm day of forty degrees Celsius, sound will travel at 355 m/s (metres per second) or 794 mph. It is worth mentioning that anyone who sucks helium into their lungs and voicebox will sound like a demented Donald Duck, as sound travels through said gas at 972 m/s or 2,174 mph.

Although movies featuring brooding dockside scenes like to give the impression that sound is muffled by fog, it in fact travels further and faster in such a medium, thanks to the increased amount of water in the air. In water itself, sound travels at over 1,500 m/s or about 3,400 mph, again depending on the temperature. In metals such as steel and aluminium, it positively races along at over 6,200 m/s, which is about 14,000 mph.

FASTER THAN LIGHT

Only within the sun can sound travel faster than light. The sun, it must be remembered, is not a solid ball but a giant sphere of searing hot plasma in which photons take over a hundred thousand years to struggle to the 'surface', or photosphere, where they are still outpaced by sound. Only when light frees itself from the sun to escape into the vacuum of space does it achieve its incredible speed of 186,000 mi./s or 299, 338 km/s.

Sound, on the other hand, travels about the surface and within the body of the sun so fast that it makes the entire mass vibrate. The hotter the plasma the faster the sound waves can travel, with speeds ranging from about 8,000 m/s or 17,895 mph to about 10,000 m/s or 22,369 mph, which is about eight times faster than light trapped in the photosphere.

DOPPLER'S SOUND AND LIGHT SHOW

In the late 1830s the Austrian physicist Christian Doppler (1803–53) – Johann Doppler is a common misnomer – was fascinated by the fact that the light from some stars presented a blue tinge while others showed reddish. He hypothesized that the blue stars were most likely to be moving towards the earth while the red ones were moving away.

His reasoning, which was quite correct as things turned out, was based on his suspicion that light waves emitted from the blue stars were being compressed tighter together due to the stars' forward movement in relation to the earth, while the redshift of the others was caused by their light waves being stretched further apart in the wake of their retreat.

In 1842 Doppler published a paper explaining how the frequency of any light wave altered the colour perceived by an observer depending on the position of that observed in relation to the speed and direction of travel of the source. Few believed him, so he decided to set up one of the most bizarre public experiments ever undertaken...

THE TRAIN NOW APPROACHING

Prior to 1845, when Christian Doppler set up his experiment, the only people aware of the compression of sound waves and the resultant alteration in pitch were soldiers who had heard bullets and cannonballs whistling past them. But they, of course, had too much on their minds at such time to jot down notes about the change in pitch of the sound emitted by that which had nearly killed them. Having obtained a general consensus from his peers that sound waves would behave in a similar way to light waves in such circumstances, Doppler hired a train and an entourage of trumpeters.

After they had checked the pitch of the instruments and agreed that all would play the note of C, half of them were installed on a flatcar behind a railway engine that would tow them past the other batch of trumpeters on a platform beside the tracks. As the train approached at the breakneck speed of forty miles per hour (64 km/h), it was clear to all present that the trumpets on approach seemed to be playing a note that was slightly higher than that played by their fellows on the platform, and that the notes only matched as the train drew level before slipping out of harmony again after the train had passed.

This proved that wave frequency, be it from light or sound, emitted by a static source, will radiate evenly like ripples from a stone thrown into a pond. But if the source is moving then the waves in front become increasingly impacted by increased speed, while those trailing behind get further and further apart. In the case of sound, this impaction of the waves raises the perceived note of the sound and, in the case of light wave impaction, moves the perceived colour further towards the blue end of the spectrum.

BREAKING THE BARRIER

Not that Doppler or any of his contemporaries could have imagined at the time, but if a sound source approached the speed of sound itself, then the sound waves ahead of that source would be standing cheek-by-jowl to create an almost physical barrier. If it has enough power to punch through that barrier and exceed the speed of sound, that barrier will 'shatter' to create a sonic boom from the shock wave.

This is why soldiers say that you never even hear the bullet that gets you, as it has already hit you before the sound arrives. Likewise, modern strike jets can wreak their havoc with missiles and be gone before anyone on the ground has even heard them coming.

WHIPCRACK AWAY!

Most maintain that the first sonic boom was generated by USAF Captain Chuck Yeager (b. 1923) flying his X-1 over the Mojave Desert on 14 October 1947.

Actually, this only made Yeager the first person to break the sound barrier: sonic booms had been heard for centuries without any realizing what they were hearing. Apart from supersonic ammunition, which has been in use for longer than you might suspect, the first person to crack a bullwhip was the one to create the first sonic boom as the tip of the lash accelerated through the sound barrier – and bullwhips have been around for perhaps two thousand years.

SPEED VERSUS VELOCITY

Although in general speech speed and velocity are synonymous, this is far from the case in scientific circles.

The measurement of speed is scalar, meaning that it does not take into account the direction of travel, and is expressed in a single figure. Velocity, on the other hand, is a figure arrived at by dividing the overall displacement of the unit in motion by the time duration of its travel.

If a runner sets out from a fixed point of reference to find himself twelve miles southeast of the starting point one hour later, then his average speed is a straightforward twelve miles per hour. His velocity, however, is twelve miles per hour southeast. If that same runner set out again to run six miles north and then back to the starting point his average speed would still be twelve miles per hour but his velocity would be zero. At the end of the allotted time he is exactly where he started so a displacement factor of zero, divided by any figure you care to mention, will always be zero.

IT'S FOR YOU!

Widely credited with having invented the telephone in 1876, Scotsman Alexander Bell (1847–1922) – 'Graham' is not on his birth certificate – actually stole the design from US electrical engineer Elisha Gray (1835–1901), who had filed for a patent at the same time that Bell was still struggling to get his own defective design to work.

The receiving clerk, Zenas Wilber, had served under Bell's lawyer, Marcellus Bailey, during the American Civil War (1861–5) and so tipped off the dodgy duo, meeting them in a hotel room to allow Bell to modify his design in accordance with Gray's submission. Wilber then took Bell's papers back to the Patent Office where he juggled the dates and records before informing Gray that his patent was denied as Bell had filed first.

Be that as it may, as Bell's telephone network expanded across America the engineers encountered the problem of the electrical current in the wires gradually losing power the further it had to travel. Bell came up with the bel/decibel scale to measure any attenuation in electrical flow in relation to the length of any wire so he could in turn work out where he had to situate electrical substations to boost the current, and thus the volume of both talking parties.

The bel and its subdivision, the decibel, were devised to measure electrical current but because this affected the sound levels of telephone lines, so many people mistook the decibel for a unit of sound measurement that it eventually became such.

CLIMBING THE SCALE

Decibels – few talk of bels these days – present on a logarithmic scale which takes the threshold of human hearing as its baseline and a scale on which, for example, twenty decibels is ten times louder than ten decibels, thirty decibels is one hundred times louder than ten decibels and two hundred decibels is one thousand times louder than one hundred decibels.

The scale starts with the threshold of human hearing marking point zero, and forty decibels up from that has you standing in a quiet residential street or a park. Move up another forty decibels and you would be standing next to a busy motorway, with a further forty decibels placing you beside a small private jet on take-off. This point of one hundred and twenty decibels is also the threshold of pain so, add another forty decibels to place you next to a space shuttle on take-off and your ears would literally start to bleed as you go deaf for life.

BRINGING THE HOUSE DOWN

Not wishing to be a party pooper but, bearing the above figures in mind, it is easy to see how regular or even a single attendance of a rock concert can seriously damage your hearing. Sound readings taken thirty metres from the stage now routinely register in the top one hundred and twenty decibels, with the heavy metal band Motörhead taking things to an unprecedented extreme in 1984, when they quite literally brought the house down in Cleveland, USA.

During the band's performance at that city's Variety Theatre, with their sustained output exceeding one hundred and thirty decibels, fans fled in panic as lumps of the ceiling started to fall on them. The band – either unwitting or uncaring of the chaos it was causing – played on until more level-headed roadies cut the power to the stage. The structural damage was subsequently assessed to be so severe that the venue was shut for good.

FOR GOOD
MEASURE

F inally, Pandora's box of units and scales that refused to sit comfortably in any other section. Many measurements considered for inclusion were screened out for being of restricted use or nebulous definition. The Garn, for example, is used by NASA as a unit on their scale of susceptibility to space sickness, an eponymous honour bestowed on Edwin Jacob Garn (b. 1932) who in 1985 became the first serving Member of Congress to go into space. Despite having been a Navy pilot with over ten thousand flying hours under his belt, he apparently vomited enthusiastically throughout his one hundred and sixty-seven hours in orbit.

Other exclusions included the mickey and the Lovelace from computer-speak. The former, an obvious nod to Mickey Mouse, denotes the smallest discernible movement of a mouse cursor, while the Lovelace scale measures user-dissatisfaction with the 'clunkiness' of a computer operating system – another eponymous honour, commemorating mathematician Ada Lovelace (1815–52), the only legitimate daughter of Lord Byron, who in the 1840s helped Charles Babbage (1791–1871) to invent the first programmable computer. That said, with information ranging from the history of clothes' sizes to the estimated calorific intake of Santa on Christmas Eve, there should be something in this section to either amuse or inform.

20/20 VISION

A measurement of visual acuity, 20/20 is popularly taken to indicate perfect eyesight, but in fact means no such thing.

If a person is rated as having 20/20 vision this only means that they can clearly see at a distance of 20 ft that which they should be able to. A person with 20/100 vision, for example, has to be standing as close as 20 ft to see what a normal person can see at 100 ft.

It has to be understood that we do not see with our eyes but with our brains, which interpret the messages sent down the optic nerve from the retina: it is the clarity of the data so transmitted that is rated on the scale. A person with perfectly normal and healthy eyes can easily have seriously impaired 'brain-vision' after a stroke or a heavy blow to the head.

SPF SCALE

A dangerously misunderstood scale of 'measurement' which does not in fact relate to anything other than individual tolerance of the prevailing levels of sunlight. Nor is the SPF scale mathematical in its escalation – SPF 15 gives 93 per cent protection; SPF 30 gives 97 per cent protection; SPF 60 renders 98.5 per cent protection – so the higher up the scale you go the smaller the difference to choose between them.

The rating actually relates to the alleged time a user can remain in the sun without burning. If you can normally stay in the direct sunlight of a given strength for about twenty minutes without any protection and without starting to burn, then repeated applications of SPF 30, for example, claim to allow you to remain in that sunlight for thirty times longer.

But that is about ten hours, which does seem risky. Alternatively, if you are more sun-resistant and able to take one hour in that same level of direct sunlight without burning, then repeated applications of SPF 30 *should* allow you to stay outside for thirty hours. Really? Providing the claims are genuine, SPF ratings only make sense when related to individual tolerance to varying levels of sunlight.

Numerous trials have failed to find any better protection from more expensive products over supermarket own-brand lotions, so the best advice is perhaps to just save your money, cover up and stay pasty-faced.

PERSONAL GROWTH

The growth spurt experienced by humans in the first two years of life is responsible for about 50 per cent of the complete cycle, so if you double a boy's height at that age you will have a good idea of his adult height. Girls develop more quickly, so the same calculation for them is made at eighteen months. Genetics also play a significant role in the determination of adult height, so another way to calculate this is to add together the heights of both parents and, in the case of a boy, add 2.5 in. / 6.35 cm and deduct the same for a girl, before dividing the remainder by two.

First established by the sculptors of ancient Greece, the human body can be measured in multiples of certain lengths and spans. The average person is, including their head, seven and a half times the length of their head, while most of the willowy and elegant prowlers of the catwalk seem to average eight times their head's length. Again, the average person, from the ground to his or her hairline, stands eighteen times the width of their fist at the knuckles, with the Naomi Campbells of this world standing at something slightly over nineteen fists.

The usual perception of the ideal European female face dictates that the distance between the pupils of the eyes should be 46 per cent of the total width of the face at the temples and that the distance from the corners of the eyes to the corners of the mouth should be 36 per cent of the total depth of the face, as taken from the hairline to the tip of the chin.

The eyes themselves should be, from the inner corners, one eye-width apart and set the same distance in from the side of the head. The length of the nose should be no more than twice the width of the eye with the philtrum of the top lip set one eye-width below the tip of the nose. The mouth itself should approximate two eye-widths. Naturally, there are many women of considerable allure who stand outside these so-called 'ideal' parameters.

For some, the perfect male form is presented by the statues of ancient Greece as defined by Polykleitos (*c.* 450–420 BC), a sculptor who laid down a table of body proportions equating everything from the length of the limbs (and yes, that as well) to multiples of the length of the distal phalange, or the top section, of the little finger. The measurement of the fist across the knuckles, for example, should be three pinkie-tips wide. Although this might seem to be taking things a tad too far, it was a proportional measurement scale that worked, and one still used by sculptors today.

THE PHYSICS OF SANTA

According to the latest figures from the Population Reference Bureau, there are about two billion children in the world. Even after deducting those of non-Christian faiths, this still leaves Santa having to deliver to about 15 per cent of that total. Allowing for 2.5 children per household that works out to 91.8 million homes.

Taking into account the various time zones and the rotation of the earth, this allows Santa a Christmas Day of about thirty-one hours during which he has to visit 822.6 houses every second. This allows him about one thousandth of a second to park on the roof, scuttle down the chimney and deliver the gift, consume his mince pie and milk and get to his next stop, averaging a speed of 650 mi./s / 1,046 k/s.

The distribution of the global Christian community gives Santa a journey of about 78 million mi. / 125 million km, during which he has to consume 91.8 million mince pies and, if every house leaves out but 200 ml (7 fl oz) of milk, he will also have to drink the equivalent of four Olympic swimming pools. This intake of 20,655,000,000 calories means Santa would put on some 2,950 tons in the course of the night, which he could work off by walking round the world twenty-eight thousand times. At least this explains what he gets up to for the rest of the year.

The average reindeer can haul about 136 kg / 300 lb and, allowing that each present weighs about a kilogram / 2.2 lb, this produces a sleighload of 321,300 tons, which would need 2,142,000 reindeer to pull it.

An object weighing 321,300 tons travelling at such speed would create enormous air resistance, subjecting the lead reindeer to fourteen quintillion joules of energy per second – little wonder they have red noses! All the reindeer would vaporize in 4.2 thousandths of a second, leaving Santa exposed to 17,500 g (g-force), which would pin him back in his seat with 4,315,015 lb / 1,957,258 kg of pressure.

Everyone in Santa's path would be killed by the sonic shock waves his journey created. The deceleration at his first stop would whiplash him and his payload into space in a fraction of a second. Happy Christmas!

WOMAN VERSUS ELEPHANT

If the woman is wearing stiletto heels she will beat the elephant every time when it comes to the load that either can exert on the surface they stand on.

If a woman weighs, say, 45 kg / 100 lb, that body weight multiplied down to the point of one of her killer heels leaves her exerting over two tons of pressure per square inch or 315,000 kgf/cm^2, which explains why such footwear is banned from all the ancient monument sites of Greece – as indeed it is from most buildings of historical interest around the world.

That 100 lb / 45 kg woman would be exerting about fifteen times the pressure per square inch of a two-ton elephant standing on just one of its broad feet. I am not suggesting for a minute that being stepped on by an elephant would be a pleasant experience, but even a sylph-like seven-stone woman standing on your bare foot would put the heel straight through it.

Now if you could get the elephant to wear stilettoes you'd *really* be in business!

THE HYNEK SCALE OF CLOSE ENCOUNTERS

Dr Josef Hynek (1910–86) was a mainstream US astronomer hired by the USAF to debunk and defuse the 1950s panic over alleged UFO sightings. But, the more reports he studied the more his opinions changed to categorize such encounters on the following scale:

Close Encounter of the First Kind: visual sighting of a UFO within 500 ft / 152 m.

Close Encounter of the Second Kind: one involving electronic disruption, physical or mental distress in those present.

Close Encounter of the Third Kind: a UFO incident in which some kind of alien life is present.

Close Encounter of the Fourth Kind: UFO incident in which human abduction occurs.

Close Encounter of the Fifth Kind: one involving human–alien communication or interaction.

Close Encounter of the Sixth Kind: one involving the death of either human or animal life.

Close Encounter of the Seventh Kind: interaction that results in the creation of an alien–human hybrid.

THE TORINO SCALE

Professor Richard P. Binzel (b. 1958) of MIT (Massachusetts Institute of Technology) devised this scale in 1995 as the Near-Earth Object Hazard Index, but renamed it after attending a conference debating the likelihood of such incidents that was held in 1999 in Turin, Italy.

Level 0: applies to objects on a certain collision course with earth but ones so small that they will burn up on entering the atmosphere.

Level 1: an object bound for our corner of space but one which will pass with no possibility of collision.

Level 2: an object that will pass close enough to the earth to warrant its being monitored.

Level 3: an object with a 1 per cent chance of collision, or marginally greater, but only large enough to cause localized destruction.

Level 4: same as Level 3 but relating to objects large enough to cause regional destruction.

Level 5: an object on a more likely impact course than at Level 4 and one large enough to cause regional devastation. Needs monitoring to update the risk of impact.

Level 6: same as Level 5 but relating to objects large enough to cause something short of global catastrophe.

Level 7: same as Level 6 but relating to objects large enough to cause global catastrophe and on a far closer impact course.

Level 8: collision certain; localized destruction if impact is terrestrial, or devastating tsunami if impact is oceanic.

Level 9: certain collision with an object larger than at Level 8, causing regional devastation or tsunamis of greater magnitude than at Level 7 if impact is oceanic.

Level 10: collision certain with an object large enough to destroy all life on the planet and possibly the planet itself.

THE LAST DROP

Frustrated at the increasing number of failed executions such as the botched hangings of criminals such as Joseph Samuel (*c.* 1780–1806), when the rope snapped not once but twice at Parramatta in Australia, and John Lee (1864–1945) who survived three attempts at Exeter prison in 1885, a committee was convened in 1886 to work out the optimum rope strength and a table of drop heights correlated to body weight.

Chaired by Henry Bruce, Baron Aberdare (1815–95), father of Charles Bruce, the leader of the Everest expedition that claimed the life of George Mallory in 1924, it was first decided that the rope employed should be able to support 1,000 lb (454 kg) for five minutes without stretching beyond a 5 per cent tolerance. Next it was worked out that the drop energy should be at least 1,260 ft-lb / 1,708.33 Nm (Newton metres) to ensure death by broken neck but without detaching the head.

This proved excessive so, after a number of decapitations by such a jolt, the drop-energy was revised to 840 ft-lb / 1,138.89 Nm, but Aberdare's meticulous scale of drop distances was, with minor adjustments, used for decades throughout the British Empire and, indeed, most of the developed world, and is still adhered to in Singapore.

Aberdare's committee laid out a table starting at 8 st / 51 kg for women and teenagers, who required a drop of 10 ft / 3 m. Moving up the weight scale in incremental steps of 7 lb / 3.18 kg, an average of 3 in. / 8 cm was deducted at each step to finish at a drop of 6 ft 5 in./ 1 m 95 cm for those weighing 14 st / 89 kg or over.

CLOTHES AND THEIR SIZES

Standardized clothing sizes are comparatively recent, with true sizing standards being set in the 1940s. Prior to this, those with money commissioned bespoke clothing, while those less fortunate either made their own or bought second-hand clothes. That is why most nineteenth-century European women of modest means got married in black. They knew their wedding dress would be the only one they could afford to have made for them and, with most men predeceasing their wives, it would later come in handy for a mourning dress. Unromantic, perhaps, but very pragmatic.

The first step to standardized measurements was forced by the American Civil War, which created such a sudden and enormous demand for uniforms that cottage industry gave way to factory production. Working to a size scale of shoulder width and height, the uniforms for the Union Army ranged from size 1 to size 10. This development did not escape the notice of publishers of mail-order clothing catalogues anxious to hit the domestic market.

The first size scale for women's clothing instituted by the catalogues ranged from size 8 to size 40, but this was not an indication of body size – size 8 was made to fit the typical girl of that age. With this ridiculous scale clearly destined for failure, and the catalogues' realization that returns were costing them over $10m a year, in 1932 they turned to the Department of Agriculture – a strange choice, but there it is – to come up with a solution.

The department hired researchers Ruth O'Brien and William Shelton to take fifty-nine body measurements from each of one and a half thousand subjects to produce a scale of twenty-seven sizes that proved too complicated to implement. Apart from that, the study was fundamentally flawed, in that all participants were white and from the lower socio-economic sector which, then as now, included a disproportionately high number of larger sizes. In 1948 the Mail Order Association of America made a wiser choice by turning to the National Bureau of Standards to institute a similar study of women serving in the US Air Force.

Although it took ten years to complete, this was the study that produced the now-familiar range of sizes 8 to 38, with overrides of T for tall, R for regular and S for short. That said, any woman who has bought clothes online or gone by the size tag in a shop will know that rarely pans out in practice. But things are set to change as online vendors – just like those early mail-order operations in nineteenth-century America – are faced with the fact that over half their sales are returned due to problems of size.

Three-dimensional digital body scans are already being promoted by firms such as Bodi.me so women can send out their measurements to calculate their size at subscribing online vendors, while eBay has set up its own bespoke service. More far-reaching than allowing people to mail-order clothes that actually fit them, this will allow Europeans to email their avatar and designs to the famously fast and remarkably cheap tailors and dressmakers of Hong Kong.

RICHTER SCALE

Devised in 1935 but now rarely used in seismological circles, Charles F. Richter's (1900–85) scale for earthquakes was a logarithmic one that measured the amplitude of the shock waves as recorded by seismographs operating many miles from the event.

Each numerical step up the Richter scale denoted an event tenfold the strength of the former, so six would denote an event ten times the strength of one registering five but a hundred times the strength of four. The US Geological Survey estimates that there are several million earthquakes around the world every year, with most occurring in areas so remote that they go unregistered. Of the five hundred thousand quakes registered on seismic equipment only one hundred thousand are felt by the local population and less than a hundred cause damage.

Only in popular speech is 'epicentre' used to mean slap-bang in the middle. In seismology the term denotes that point on the earth's surface which is directly above the fault-rupture but, as that epicentre could be miles above that point and with fault lines often slanting gradually up to the surface, the impact of the quake could be hundreds of miles away, while those standing above the epicentre notice nothing at all.

CAR ENGINE RATING

James Watt (1736–1819) met with resistance when trying to sell his steam engines to the mining industry, which used shaft pumps, powered by horses, to keep the network of tunnels free of flooding. To give himself a sales gimmick, Watt worked out that the average workhorse, attached to a pulley, could lift 550 lb / 250 kg through 1 (vertical) ft / 30 (vertical) cm in one second, so he rated his new engines by horsepower to give his equine-fixated customers some idea what was on offer.

With the first viable petrol-powered cars having been developed in Germany, France and Italy, engine capacity is traditionally given in litres or cubic centimetres, which is perhaps for the best as a seven-pint Smith doesn't sound as sexy as a four-litre Ferrari (the Italian equivalent of Smith, a name denoting a man who works with *fer*, or iron).

THE GOLDEN RATIO

There is a little-understood mechanism in the human brain which finds shapes presenting what is called the golden ratio of 1:1.618 more pleasing than any other. Perhaps it is nothing other than the narcissistic pleasure we derive from looking at ourselves, as the human face is a manifestation of the ratio which also pervades nature, to be seen in everything from pinecones to the patterns of hurricanes and the spirals of the galaxies.

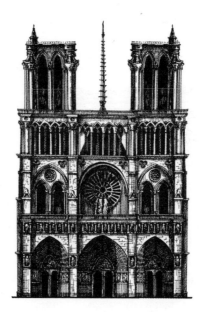

Notre-Dame Cathedral, Paris

The human eye will recognize a rectangle or any other pattern presenting such measurement faster than it will any other shape, which is why architects have used it in structures ranging from Stonehenge and the Parthenon to the Great Mosque of Kairouan and the façade of Notre-Dame.

Artists have always used it to define the dimensions of their finished work and it is even built into Western musical structure so we can 'hear' it. Most of the books you buy manifest a 1:1.618 format, as does the credit or debit card you use to buy them.

UNUSUAL MEASUREMENTS

BED – banana equivalent dose. Believe it or not, the humble banana is naturally radioactive, so its emission-level is used to gauge low-level or background radiation. Don't worry, it would take thirty-five million BEDs to kill you.

BMI – Big Mac Index. As has the Mars Bar been used in the UK to measure the relationship between wages and prices, the fairly static price of the Big Mac is employed in the USA to see, for example, how many Big Macs equate to the price of a gallon of petrol today as compared to ten years ago.

CAR – a measure of 4 m / 13 ft, this is used to give braking distances.

COW-GRASS – an agricultural term for the area of pasture required to support one cow.

FINGER – a rough US drinks measure equating to the depth indicated by a number of fingers held against the side of the glass – two fingers of rye, for example.

HIROSHIMA – a measure used by geologists to describe the energy-release of earthquakes and volcanic eruptions, or by astrologists for asteroid strikes with one Hiroshima equivalent to fifteen kilotons of TNT. Prior to this, the Halifax (3 kt TNT) was used, after a 1917 munitions explosion in that Nova Scotia town that flung a half-ton ship's anchor four kilometres, or two and a half miles, into the neighbouring town of Armdale.

JIFFY – yes, it does exist, as a measure of 0.01 of a second.

LENGTH – mainly heard at the racetrack, this was originally a horse-length, or 8 ft / 2 m.

LD – or lunar-distance from earth, being 380,000 km / 236,000 mi.; this is used by space-agencies to define near misses of objects passing our planet.

MCG – Melbourne Cricket Ground, an Australian measure used to denote any large number of people, with one MCG indicating 95,000 – the seating capacity of said stadium.

OSP – Olympic swimming pool. If used of area this is 50 m by 25 m / 164 ft by 82 ft. If expressing volume, it is 550,000 imperial / 600,000 US gallons.

PONY – in the UK this is three-quarters of a fluid ounce (21 ml), but one fluid ounce (30 ml) in the USA.

SYDARB – Sydney Harbour, an Australian measure of large volumes of water, each sydarb being 0.5 km^3 / 500,000,000 m^3.

Sydney Opera House

WALES – an area unit equating to the size of that country, 20,779 km^2 / 8,023 mi^2, and one used by rainforest conservationists. In the USA the Rhode Island is a much-used measure of 4,000 km^2 / 1,545 mi^2, while the CIA speaks in terms of Washingtons (DC), 159 km^2 / 61.4 mi^2. In Russia they discuss large areas in terms of the France, 551,695 km^2 / 213,011 mi^2.

SELECT BIBLIOGRAPHY

Darton, Mike, and Clarke, John, *The Dent Dictionary of Measurement* (J. M. Dent, 1994)

Duncan, David Ewing, *The Calendar: The 5000-Year Struggle to Align the Clock and the Heavens, and What Happened to the Missing Ten Days* (Fourth Estate, 1998)

Kemp, Peter (ed.), *The Oxford Companion to Ships and the Sea* (Oxford University Press, 1976)

Muir, Hazel, *Eureka!: Science's Greatest Thinkers and Their Key Breakthroughs* (Quercus, 2012)

Smith, Timothy Paul, *How Big is Big and How Small is Small: The Sizes of Everything and Why* (Oxford University Press, 2013)

Strauss, Stephen, *The Sizesaurus: From Hectares to Decibels to Calories, a Witty Compendium of Measurements* (Kodansha America Inc., 1995)

Whitelaw, Ian, *A Measure of All Things: The Story of Man and Measurement* (Quid Publishing, 2007)

PICTURE
ACKNOWLEDGEMENTS

Height, Length and Depth p.12 © Lestyan (www.shutterstock.com); p.13 © Emre Tarimcioglu (www.shutterstock.com); p.16 © Jka (www.shutterstock.com); p.20 © Mary Evans Picture Library; p.23 © Clipart.com; p.32 © Everett Historical (www.shutterstock.com); p.34 © Pim (www.shutterstock.com)

Distance p.43 © Clipart.com; p.47 © Elena Terletskaya (www.shutterstock.com); p.54 © Morphart Creation (www.shutterstock.com);

Volume and Area p.62 © Sharpner (www.shutterstock.com); p.66 © Mary Evans Picture Library; p.68 © Vlada Young (www.shutterstock.com)

Weight, Displacement and Density p.76 © MaKars (www.shutterstock.com); p.89 © Clipart.com

Dates and Calendars p.97 © Clipart.com; p.99 © Mary Evans Picture Library; p.106 © Argus (www.shutterstock.com)

PICTURE ACKNOWLEDGEMENTS

Time and Clocks p.115 © Clipart.com; p.119 © Clipart.com; p.124 © Pim (www.shutterstock.com); p.126 © Clipart.com; p.133 © iStock

Temperature p.139 © Clipart.com; p.144 © Clipart.com; p.149 © Sketch Master (www.shutterstock.com)

Sound and Light p.152 © HiSunnySky (www.shutterstock.com); p.161 © Ad libitum (www.shutterstock.com)

For Good Measure p.167 © iStock; p.174 © Alex74 (www.shutterstock.com); p.178 © AVA Bitter (www.shutterstock.com); p.181 © Akvaartist (www.shutterstock.com)

INDEX